AFRICAN ISSUES

**Electricity
in Africa**

AFRICAN ISSUES

AFRICAN ISSUES

Electricity in Africa

The Politics of Transformation in Uganda

CHRISTOPHER D. GORE

JC JAMES CURREY

James Currey
is an imprint of
Boydell & Brewer Ltd
PO Box 9, Woodbridge
Suffolk IP12 3DF (GB)
www.jamescurrey.com
and of
Boydell & Brewer Inc.
668 Mt Hope Avenue
Rochester, NY 14620-2731 (US)
www.boydellandbrewer.com

British Library Cataloguing in Publication Data
A catalogue record for this book is available on request from the British Library

ISBN 978-1-84701-168-8 (James Currey paper)
ISBN 978-1-84701-169-5 (James Currey cloth)

This book is printed on acid-free paper

Typeset in 9/11 Melior with Optima display
by Avocet Typeset, Somerton, Somerset TA11 6RT

Printed and bound in Great Britain by
TJ International Ltd, Padstow, Cornwall

MIX
Paper from
responsible sources
FSC
www.fsc.org FSC® C013056

DEDICATION

This book is dedicated to all those working to improve access to modern, sustainable energy in Uganda, and the citizens, particularly women and children, waiting for clean, modern energy to arrive.

CONTENTS

5

**Electricity
& the Politics
of Transformation**   145

LIST OF ILLUSTRATIONS

ACKNOWLEDGEMENTS

This book is a product of a great deal of institutional, family, and peer support, both in Canada and in East Africa. My heartfelt thanks go to the many people in Uganda and Kenya who have shared their knowledge, time, stories, and reflections with me over many years. This book would not have been possible without the patience of the many individuals and organizations willing to speak with me and to inform me of events, debates, and challenges in the region. I would particularly like to thank the many civil servants from various government and quasi-government agencies who spoke with me and patiently answered my questions, particularly in the Ministries of Energy and Minerals Development; the Rural Electrification Agency, and the National Environmental Management Authority. The subject of this book was and remains controversial. As a result, I am very thankful for the honest and open reflections offered by many people. I have tried to present their perspectives as clearly and fairly as possible. Similarly, I appreciate the candid perspectives offered by representatives of private energy firms and the many representatives of bilateral and multilateral agencies who met with me while in Uganda. All demonstrated a strong interest in explaining the difficult challenges they encountered in the country.

Given the controversial aspects of this research subject, I also thank the many non-government organizations in Uganda who met and spoke with me. Despite their lack of resources, all were keen to discuss their work and to share their knowledge. Many individuals and organizations deserve special thanks, including the National Association of Professional Environmentalists (NAPE); the Advocates Coalition for Development and Environment (ACODE); and International Rivers. I also owe special thanks to the Centre for Basic Research in Kampala for hosting me many years ago when I started this research, and to the late Executive Director, Dr Bazaara Nyangabyaki.

I am also very thankful to the many citizens who spoke with me informally about Uganda, and about their day-to-day challenges. My efforts to explain the link between national and international policy prescriptions and 'life on the ground' is certainly informed and motivated by the numerous, formal and informal, friendly, and passionate conversations I had in the houses, on the streets, and in the buses of Uganda. I hope the book is richer for having tried to integrate these perspectives into my work and to connect the micro- and macro-level politics surrounding energy. I am also grateful to Dr Suzie Muwanga for her continued assistance and support, and for providing me with a 'home away from home' while in Uganda. Her friendship, support, and advice have been invaluable and I am indebted to her and her family for welcoming me into their lives.

In Kenya, I want to thank Davinder Lamba, Diana Lee-Smith, and Stephen Karakezi. During my first trip to East Africa, Mr Lamba advised me and provided me with quiet sanctuary at the Mazingira Institute while I considered how to approach the complex issue of energy. Since that time, Davinder and Diana, co-founders of the Mazingira Institute, have served as friends and mentors. I'm indebted to them for their guidance and support. Stephen Karakezi, and his colleagues at the African Energy Policy Research Network (AFREPREN), also deserve special thanks. During my first trip to Nairobi, Mr Karakezi spoke with me for a great length of time, sharing his extensive knowledge of energy issues in East Africa. Mr Karakezi remained an invaluable colleague, whose long-term commitment to energy issues in Africa, particularly energy services to the poor, is infectious and inspiring.

At the University of Toronto, my thanks to the members of my PhD thesis committee who, nearly twenty years ago, believed that a study of the politics of electricity in Africa was a worthwhile endeavour: Professor Dickson Eyoh, Professor Rodney White, Professor Patricia McCarney, and Professor Richard Stren. Professors Eyoh and White provided sage advice and constructive feedback throughout my research. I'm sad that Professor White is not alive to see the publication of this book: he challenged me to think critically about energy issues, particularly by asking whether 'big problems, require big solutions'. Professor Patricia McCarney was one of the first faculty members I met at the University of Toronto as a graduate student. Her enthusiastic and constant support over many years has been tremendous. She instilled in me the need to carefully understand the relationship between international programmes and policies and on-the-ground needs and realities, and the importance of bringing these two realities to light in research and practice. I am enormously grateful to my PhD thesis supervisor, Professor Richard Stren. Professor Stren is one of the most generous individuals I know and a model for the type of scholar and teacher I have aimed to become. Professor Stren reminded me that above anything else, the most important thing was that my research be of value and importance to the region and country in which I am studying. With this priority established, the academic rigour would follow.

Indeed, when I first travelled to East Africa in the early 2000s, I was interested in studying the transboundary governance of Lake Victoria. But it was one meeting with a senior bureaucrat in the National Environmental Management Authority that changed my research path. He said, 'Lake Victoria is interesting, but if you want to study a real challenge in Uganda, study electricity!' And with that, further conversations with citizens through to donors revealed the scale of the challenge in the sector. Professor Stren encouraged me to examine electricity when there were few examples of how this might be done. His mentorship and friendship have had a lasting impact on my scholarship, particularly on my goals to collaborate with researchers in East Africa, and to thoughtfully and critically examine political challenges through the lens of residents through to international donors.

I started this research in 2000. Having moved from a graduate student to a faculty member during this time, I've been fortunate to receive support from the Faculty of Arts, Ryerson University, and the Department of Politics and Public Administration. In recent years, I have also been very fortunate to connect and now work with several other people equally fascinated by the political challenge of energy and electricity. Three colleagues and friends deserve very deep thanks: Drs Lauren MacLean, Jennifer Brass, and Elizabeth Baldwin. They reinvigorated in me a desire to connect my early studies of electricity to contemporary events and to share this knowledge in a book. I also want to thank Drs Harriet Bulkeley, Marcus Power, and Peter Newell for inviting me to participate in discussions and debates about Rising Powers and inspiring me to think more carefully about the politics of transitions and transformations.

I also want to sincerely thank Douglas Johnson and Jaqueline Mitchell at James Currey. Douglas was the original Commissioning Editor at James Currey who encouraged me to complete this manuscript before retiring. Jaqueline carried on Douglas' encouragement, and her patience and guidance through this publishing process has been most welcome and appreciated.

Finally, my sincerest thanks go to my immediate and extended family. My mother and father, Sheila and David Gore, my brother, Michael Gore, and sister-in-law Laureen Gore, never questioned the rationale or motive for the work I was doing, and have provided direct and indirect support to me as I continued to study and work in East Africa over the last fifteen years. Their confidence in me, and their love and support has been enormously valuable and I am tremendously grateful for it. I also thank my extended family, Beth and Clark Carnegie and Bruce and Anita Robinson for their ongoing support, and my dear friends, Dr Konrad Ng and Luke Ferris, for helping me to see the lighter side of life amidst the challenges of academic life.

Lastly, but most importantly, I thank my wife and dearest friend, Dr Pamela Robinson. Her love and support since I first travelled to East

Africa, through to the dozens of trips since, has been unwavering. My research is stronger for having the opportunity to share and discuss it with her, and my life is richer for having the chance to share it with her. I also thank my two daughters, Ruby and Hilary, who bring joy to my days while also challenging and inspiring me. I'm grateful to have had the chance to share East Africa with them, and hope they have the fortune to pursue work and build global relationships that inspire them too.

Some historical context and quotes in this book originally appeared in Gore, C. (2009), 'Electricity and privatisation in Uganda: The origins of the crisis and problems with the response', in D. A. McDonald (ed.), *Electric Capitalism. Recolonising Africa on the Power Grid*, pp. 359–399; Cape Town: HSRC Press. The work is reproduced with permission from HSRC Press.

Two maps used in this book were originally published in Wilson, G. (1967), *Owen Falls: Electricity in a Developing Country*, Nairobi, Kenya: East African Publishing House. The publisher was not able to be reached for copyright approval and no longer exists.

ACRONYMS

i

AESNP	AES Nile Power
EA	Environmental Assessment
ERA	Electricity Regulatory Authority
ESMAP	Energy Sector Management Assistance Program
GET FiT	Global Energy Transfer Feed-in Tariff
GoU	Government of Uganda
IA	Implementation Agreement
IDA	International Development Association
IFC	International Finance Corporation
IMF	International Monetary Fund
IPP	Independent Power Producer
KPLC	Kenya Power and Lighting Company
MEMD	Ministry of Energy and Minerals Developments
MFPED	Ministry of Finance Planning and Economic Development
MIGA	Multilateral Investment Guarantee Agency
MW	Megawatt
MWLE	Ministry of Water Lands and Environment
NEMA	National Environmental Management Authority
PPA	Power Purchase Agreement
PRG	Partial Risk Guarantee
REA	Rural Electrification Agency
UBOS	Uganda Bureau of Statistics
UEB	Uganda Electricity Board
UEDCL	Uganda Electricity Distribution Company Limited
UEGCL	Uganda Electricity Generation Company Limited
UETCL	Uganda Electricity Transmission Company Limited
UREA	Uganda Renewable Energy Association

Introduction

Uganda, in the marrow of tropical Africa, may become one of the world's greatest exporters of – electricity.

– John Gunther, 1955

On 24 January 2002, crowds gathered for a celebration on the banks of the Nile River. Just north of the town of Jinja, in the east African country of Uganda, diplomats, bilateral and multilateral agency representatives, citizens, Members of Parliament, and the President of Uganda all congregated for what was thought to be the beginning of the construction of a new 250-megawatt (MW) hydroelectric dam – the Bujagali dam.

The location of the ground-breaking ceremony held historical significance. In 1907, a young Winston Churchill, then Parliamentary Under-Secretary of State for Colonies, stood about ten kilometres downstream and reflected on the potential for Uganda to become an industrial force if the waters of the Nile could be harnessed for electric power (Churchill 1989 [1908]). By 1954, the colonial authority saw the first phase of its Ugandan hydroelectric-vision fulfilled when the Owen Falls Dam (now named the Nalubaale dam) was inaugurated in a ceremony presided over by Queen Elizabeth II. The vision of Uganda becoming a regional electricity superpower has not yet materialized, however. The Nalubaale dam remained the only major source of electricity in the country until 2012. Six decades later, despite years of effort to improve electricity access and to reform the sector, Uganda had the unenviable reputation as having one of the lowest levels of access to electricity in the world (IEA 2011). Accordingly, the inauguration ceremony for the Bujagali dam in 2002 was supposed to represent a turning point in what had been a painfully long period of poor and unreliable access to electricity in the country. Between 1971 and 1979, the period of Idi Amin's reign in Uganda, the number of electricity consumers dropped from 69,500 to 60,950 (Uganda Electricity Board 1996; 1999). During the height of civil conflict in Uganda (1979 to 1986) the number

1

of individual consumers increased to over 105,000, but two years after President Museveni took power in 1986, the country had only 80,795 consumers.

President Yoweri Museveni was the last of the dignitaries to speak at the inauguration ceremony. While other presenters praised Museveni's leadership, persistence and vision in executing the dam, Museveni had sat sternly, showing little emotion. It turned out he was in no mood for celebration or praise.

The President began by stating that he was 'not happy at all' and was 'ashamed'. He did not want to talk about how happy Ugandans were but just wanted to get on with the project. He suggested that a project that should have taken two years to launch had – by this point – taken seven. Museveni said he didn't accept any person's thanks; people were foolish for thanking him. 'Do you thank people for feeding their children?' he asked rhetorically (Okwello 2002). After describing Uganda's great potential for producing thousands of megawatts of electricity by harnessing the power of the Nile River and noting the serious electricity deficit in the country, he stated that the process leading to the construction of Bujagali was a 'circus', which had led to embarrassment and undermined their interests: 'This is an occasion of shame and repentance,' he said. Museveni finished his speech criticizing the one institution that had invested more in Uganda's electricity sector than any other: the World Bank. The first project the World Bank financed in Uganda was just prior to the country's independence and was focused on electricity: the *Electric Power Development Project* (Power I). This project was to support the Uganda Electricity Board's $14.0 million expansion programme, of which the Bank loaned $8.4 million.

The Bank 'needs to stop listening to so many people' and instead 'talk to people in the Third World,' Museveni said. The Bank 'listens to a lot of nonsense' and is 'too squeamish and too sensitive to shallow opinions of those who aren't supportive of transformation.' Museveni was making a not-so-veiled reference to Members of Parliament and domestic and international non-government organizations that had raised concerns with the dam and halted its progress through appeals to international and domestic institutions – individuals and organizations he labelled as 'economic saboteurs' and 'enemies of the state'. Two days after the ceremony, Museveni went further: 'Those who delay industrial projects are enemies and … I am going to open war on them' (Okwello 2002). The ceremony ended with Museveni climbing on to a bulldozer and demonstrating his prowess at moving the earth.

Despite this January 2002 groundbreaking ceremony, the physical construction of the Bujagali dam was delayed repeatedly and did not begin to generate electricity for another ten years. The project was originally conceived as a private initiative under a 'build-own-operate-transfer (BOOT)' arrangement, which was in keeping with a continental trend to unbundle government electricity monopolies and promote

private-led infrastructure development. In August 2003, however, estimating a financial loss of US$75 million and amidst continuing delays
in the start of construction, US-based firm AES, which had been given
the dam site by the President without a competitive bidding process,
withdrew from its protracted ten-year effort to construct the dam (see
Gore 2009). At the same time that AES withdrew from Uganda, it also
suspended a $2.5 billion investment in thermal electric power facilities
in Brazil. Despite this major setback and ongoing local and international
concern over other matters such as the political and economic risk of
the investment, the future price of electricity, alternative generation
sources, environmental impacts, resettlement, cultural and tourism
significance, lack of competitive bidding, and low water levels in Lake
Victoria, the Government of Uganda (GOU) continued in its determina-
tion to construct the dam.

For supporters of the dam, Bujagali was viewed as the least-cost
long-term solution to solve Uganda's electricity problems: 'a no-brainer',
according to then Uganda World Bank country manager (Robert Blake,
World Bank Country Manager, interview, 5 May 2002). Indeed, owing
to the comparatively few number of people that would have to be
resettled, the high banks of the river, which could be used to support
the construction, and the presence of an island in the middle of the
dam location, which would facilitate the redirection of water during
construction, a site engineer explained: 'If you want to build a dam, this
is the ideal site.' Hence, in early 2004, the Government issued a call for
tenders to resume construction of the dam. One year later, in May 2005,
the government announced that the Industrial Promotion Services
(IPS), part of the Aga Khan Fund for Economic Development (AKFED) –
the economic development arm of the Aga Khan Development Network
(AKDN) – along with its partner company, SG Bujagali Holdings Ltd, an
affiliate of US-based Sithe Global, LLC, had won the contract to finish
construction of the dam (Daily Nation 2005). The Bujagali dam was
finally operational in February 2012, two decades after formal project
implementation activities had begun and almost a century since the
dam's location had been identified as a preferred hydroelectric site by
colonial authorities. During the time it took to execute the project, the
consequences of poor access to electricity in the country proved debil-
itating domestically. It also had significant effects regionally and politically.

Electricity in Uganda: A crisis

In 2005, while Uganda's population had increased to nearly 27 million,
the percentage of people with access to electricity remained at about
4%. Adding to this, owing to low water levels in Lake Victoria, variously attributed to drought, excessive irrigation, and overuse of water

for electricity generation, in 2006 Uganda's capacity to generate electricity dropped from an estimated 500 megawatts (MW) to just 135 MW for the entire country. According to real and estimated data from the United Nations (data.un.org), in 2006, the total installed generating capacity of Uganda relative to other African countries was: Chad, 31 MW; Central African Republic, 41 MW; Rwanda, 57 MW; Senegal, 488 MW; *Uganda, 506 MW*; Ethiopia, 816 MW; Tanzania, 957 MW; Kenya, 1392 MW; Côte d'Ivoire, 1499 MW; Ghana, 1730 MW; Zambia, 1770 MW; Zimbabwe, 2005 MW; DRC, 2444 MW; South Africa, 42 500 MW. Meanwhile, 'effective demand' for electricity in Uganda – what consumers could and would pay for – was growing at about 30 MW per year (over 20% per year). In order to meet demand and limit the already routine electricity outages, two large and expensive 50 MW diesel generators were added to the national electricity grid, and electricity was imported from Kenya. The importation of electricity from Kenya was a reversal in regional electricity planning.

In the early 1990s, when there was optimism about the quick completion of the Bujagali dam, plans were already being made to sell surplus Ugandan electricity to Kenya, Tanzania and Rwanda, and for Uganda to become a regional electricity exporter, particularly as the country built more dams on the Nile. The Bujagali dam was supposed to satisfy domestic demand at the same time as serve as a dominant industrial development strategy. The failure to execute this vision resulted in neighbouring countries quickly reassessing and altering their own energy system planning, and other countries quickly trumping Uganda's aspirations as a dominant regional electricity exporter.

Kenya, for example, began major investments in new sources – geothermal, wind and co-generation – at the same time that Bujagali was delayed in the early 2000s. Ethiopia continued with massive investments in new large, controversial hydroelectric projects. For example, in 2013, a new transmission line was approved for construction between Ethiopia and Kenya, which would bring power from the controversial Gibe III dam project in Ethiopia to Kenya. Critics suggest that the transmission line, financed by the African Development Bank and the World Bank, is tacit, if not direct, support for the 1870 MW dam project that threatens to reduce water flowing into Lake Turkana in northern Kenya that will eventually lead to it drying up, undermining the livelihoods of the population in northern Kenya relying on the lake (see International Rivers 2013; Human Rights Watch 2015). In Ethiopia's rapid construction of several large dams it has secured electricity export agreements with Djibouti, Sudan, South Sudan and, most controversially, with Kenya, which is conflicted about concerns for the livelihoods of the Turkana people and its own need for electricity. The Grand Ethiopian Renaissance Dam (GERD) is another major undertaking in Ethiopia. The project, one of the largest potential dams ever built in the world, has raised significant international concern and tension in the Nile Basin owing to fears in Egypt of decreased water supplies, and international advocacy and information campaigns about the risks of the project, particularly by the international civil society organization, International Rivers. International Rivers has been critiqued harshly for being 'anti-development' – a common refrain used by governments to discredit environmental advocacy groups – when it shares expert, and sometimes leaked assessments of major dams, like the GERD project (see International Rivers 2014). Despite this, Ethiopia has used innovative domestic financing schemes to maintain public and financial support for the project. For example, it has held six rounds of domestic lotteries since 2007 to raise funds to support dam construction. In 2007, it raised 80 million birr, or nearly US$9 million, using a text messaging lottery competition (Jemaneh 2016).

Meanwhile, as Uganda awaited the construction of the Bujagali dam, rolling blackouts continued and demand for biomass for energy (firewood and charcoal), already the primary source of energy for 95% of Ugandans, rose. The situation in the country was so serious in the early 2000s that projections of 7% economic growth were reduced to 4.5% largely due to power shortages (Among & Kalinaki 2006). The capital city, Kampala, became sarcastically described as 'generator city' given the constant hum of small diesel generators (Onyango-Obbo 2006). Adding insult to injury, in late 2006, Ugandans were paying more for electricity than any other country in the region. Household consumers were paying a *subsidized* per unit price of electricity of US24 cents/ kwh. A main reason for this was the country's dependency on several, large, expensive, diesel generators to feed the national grid. The Uganda Transmission Company Ltd, owned and operated by the government, subsidized the price of electricity by US$126 million in 2006 (*Monitor* 2006a). Without the subsidy, at the end of 2006 domestic consumers would have been paying just over US 30 cents/kwh. In comparison, in the same year, among Organization for Economic Cooperation and Development (OECD) countries, the lowest average household price of electricity was US9.4 cents/kwh (Norway), while the highest was 25.8 (Netherlands); in 2005, the average per unit price of electricity among all OECD countries was 12.7 cents/kwh (IEA 2007, II.48). It was not surprising, then, that in summarizing the state of the electricity sector in Uganda in the mid-2000s, the Minister of Energy, Daudi Migereko, concluded frankly: 'We are in a crisis' (Iziwa 2006).

How did a country held up as a 'showcase' of reform (see Dijkstra & van Donge 2001), a country receiving more favourable support from donors than other neighbouring countries with similarly questionable political regimes (Dijkstra & van Donge 2001; Harrison 2001; Muhumuza 2002; Tripp; 2004, 2010), find itself in such a perilous situation and unable to execute reforms necessary to improve its energy sector and electricity access? Given that no country has developed beyond a subsistence economy without ensuring at least minimum access to electricity for a broad section of its population (World Bank 2000a), the 'electricity problem' in Uganda and sub-Saharan Africa more broadly remains puzzling. In 2014, the most dominant countries economically in the sub-continent, South Africa and Nigeria, were each embroiled in desperate efforts to maintain regular electricity supplies. In 2015 and 2016, countries throughout the sub-continent struggled to meet household and industrial demand for electricity. How can an issue deemed to be so essential to social and economic development be so difficult to address?

This book examines the politics of electricity and infrastructure provision in sub-Saharan Africa. More directly, the book examines the politics of electricity transitions and transformations, with a specific emphasis on one country – Uganda. In recent years, a great deal of exciting scholarship has started to critically examine 'energy transitions' in sub-Saharan Africa. This work combines insights about technology transitions with questions about energy and climate justice through a political economy lens (see Baker, Newell & Phillips 2014; Newell & Bulkeley 2016; Newell & Mulvaney 2013; Newell & Phillips 2016; Power et al. 2016; Scoones, Leach & Newell 2015a). As Newell and Phillips explain (2016, p. 47), 'a political economy account of energy transitions ... focuses on institutions and relations of power' to understand 'the structures and actors that govern energy regimes and the uneven outcomes they produce'. The research for this book, which started in the early 2000s, responds to the fact that to date there has been very little written about the politics of energy in sub-Saharan Africa (Newell & Phillips 2016, p. 40) or about the politics and political economy of energy transitions generally (Baker, Newell & Phillips 2014, p. 7). The book emphasizes the questions raised by scholars using a political economy approach, but gives prominent attention to the role of African national governments in these transitions. Hence, the book offers an examination of the historic and contemporary *political* challenge of providing a critical public service – electricity – explaining how this challenge has evolved in sub-Saharan Africa since the late 1990s, and documenting how energy has been governed in sub-Saharan Africa over time.

The book reveals that electricity offers a unique window into the changing political landscape in sub-Saharan Africa in the late 1990s, particularly in relation to the character of relations between national governments, citizens, civil society organizations, private firms, and bilateral and multilateral donors. The book shows that the politics and processes surrounding electricity infrastructure, provision and reform must be understood as more than conflict-laden development initiatives. The process of electricity sector reform and decisions about service provision and technological choice during the late 1990s and early 2000s, in many countries, reveals the shifting power relationships between national governments and customary development partners, and nation-states and domestic civil society organizations.[1] The efforts to transform electricity sectors and to increase electricity access are replete with tensions and conflicts at multiple scales and between multiple actors. The tensions are also historically contingent: electricity access in sub-Saharan Africa was shaped by colonial forces (Njoh 2016), and early post-colonial African leaders used electricity and infrastructure differently to help shape national politics (MacLean et al. 2017.). Hence, the history of electricity provides critical insights into present-day access but also the evolution of domestic and multilevel politics in sub-Saharan African countries.

The book follows in the spirit of Albert O. Hirschman's classic work on the politics of development projects (1967) by examining the development apparatus itself and looking at the intersection and convergence of local, national and international interests in political and development decisions. Thus, it dives into the messy processes of political and electricity transformations, and embraces the fact that 'all transformations are replete with governance challenges', which require investigators to ask 'whose rules rule, which institutions define visions of change and the terms of change, and which relations of power shape different pathways?' (Scoones, Newell & Leach 2015b, p. 6). The book adds to critical contemporary studies of how technical solutions to economic and poverty problems have often trumped the complex character of political, social and economic relations in a given country in order to achieve some vague goal of 'national development' (see Easterly 2013).

The book draws from over fifteen years of research in eastern Africa (including Ethiopia, Kenya and Tanzania), but particularly one electricity-poor country – Uganda. The research began in 2001 as my doctoral research, with eight months of field research between 2002 and 2003. Subsequent field research took place in 2008, 2010, 2012, 2013, 2014, 2015 and 2016. In total, over 180 in-depth, semi-structured interviews with government, non-government, private sector, bilateral and multilateral representatives, along with households and citizens inform the research. The research reveals the changing nature of state/non-state relations in sub-Saharan Africa and the role that large 'development undertakings' play in the development apparatus itself. Further, the research and book highlight the political and policy complexity of trying to produce an *electricity transformation* in the sub-continent and how and why certain energy pathways were chosen. As Scoones, Newell and Leach argue: 'An understanding of politics is important in explaining which pathways get supported and legitimized, and which are ignored and so fail to gain traction' (2015b, p. 7). A key question when examining the politics of transformations is 'who steers, and which

[1] Andrew Mertha's 2008 book, *China's Water Warriors*, offers an impressive account of how grassroots opposition to hydroelectric dams in China in combination with a changing political and policymaking landscape in the country produced a situation where non-state interests emerged to hold new influence in national policymaking.

actors and institutions govern transformations', owing to the fact that transformations come about through multiple potential mechanisms – technology, market, state-led and citizen-led (ibid). This is particularly important when studying electricity generally and sub-Saharan Africa specifically, because electricity has historically been a state-led service. Yet as recent history and the central case in this book highlights, one of the fundamental tensions remains what interventions are chosen and implemented, and what role the private sector, market incentives, the state, and citizens play in these transformations. While a country's end goal may be to increase access to electricity, conflicts arise over several dimensions of electricity provision: solar panels may provide much-needed light at the household level, but their promotion and uptake can be undermined by public desire for grid-based electricity that is thought to be more socially and economically transformational; hydroelectricity may be deemed to be a clean energy source by some, but if the outcome of a massive investment in hydropower primarily provides increased, reliable power to middle-income, commercial and industrial users, and the poor continue to use biomass, how are the environmental costs and benefits being calculated and disclosed, and who is involved in decisions about future electricity access? These are fundamentally questions about energy justice.

While there are a multitude of ways to consider energy through a justice lens (see Sovacool & Dworkin 2014) the two most simple aspects relate to distributional and procedural justice (see Eames & Hunt 2013). Distributional energy justice raises questions about who gets access to electricity and what form of electricity households get; that is, is it just if some have grid-based electricity and others have a small solar panel, and what is the timeline for provision? Procedural energy justice raises questions about who gets to participate in decisions about access and what criteria are used for deciding who gets access. As demand for electricity outpaces supply, researchers and energy planners must confront these challenging questions given that citizens are certainly asking them.

Using Uganda as a central case, the book makes two central contributions to understanding the politics of energy transformations in Africa generally and Uganda specifically. First, the book explains, in detail, how the World Bank as an institution – a complex organization both governed by its own rules, which dictate how it acts, while also promoting rules for African countries – and its reform and privatization agenda in the early 1990s became embedded in Uganda's energy sector. While the World Bank's involvement may not be a surprise, to date, there is little evidence of how the World Bank became embedded in the governance of energy in Uganda, and what conflicts and contests emerged from this process. Hence, the book offers insight into the domestic and multilevel politics of energy reform where privatization and large dam construction converged under the clear promotion and advocacy of the World Bank. These findings provide a detailed and multi-year understanding about how the World Bank influenced the domestic politics of the Ugandan state in the energy sector – details still not easily found in reviews of national energy reform in Africa.

The second, more compelling and novel finding is what the outcome of electricity reforms have been for national politics in and the political economy of Uganda, as well as for future energy access and interventions. The book shows that one of the reasons that the energy reform agenda encountered so many challenges in Uganda relates to the World Bank and Ugandan national government not accounting for the changing domestic political context in which reforms were being implemented. Civil society groups in Uganda were challenging the national government using more sophisticated arguments and techniques, including successfully appealing directly to formal national and international institutions. Courts and parliamentary committees were also putting checks on government decisions. Indeed, I argue that the electricity reforms enlivened the capacity and advocacy of domestic state and non-state organizations. But perhaps what is most interesting is how the Government of Uganda's own approach to electricity provision changed due to the problems it encountered in the earlier phases of reform. Given its deep frustration with the processes promoted by the World Bank, and the failure of those processes to produce the desired improvements in electricity supply, the Government of Uganda made an explicit choice to move away from its 'traditional' lending partners and to embrace new approaches and partners for electricity. Chief among these new partners was China. Hence, the effect of the privatization and reform agenda and process promoted by the World Bank has been a realigning of donor–state relations in the country, with Uganda moving away from the World Bank, turning inward to strengthen domestic control over energy decisions, and seeking financial and infrastructure building support from new partners, particularly China. In effect, there have been *multiple transformations* in Uganda resulting from its experience of trying to improve its electricity sector: technical, social, political and economic. While researchers have clearly recognized that 'multiple forms of energy transition are being orchestrated by a range of actors within beyond the state' (Newell & Bulkeley 2016, p. 7), it is important to distinguish between technical energy *transitions* and the multiple, contested political and social *transformations* that underpin those transitions. This is the goal in this book: to provide a detailed account of the multiple and reciprocal ways that politics and technical energy goals interact. While Uganda's experience is central in this book and its experience is a central focus, the country's struggle with electricity is not unique in the sub-continent where poor electricity access and blackouts remain common.

Organization of book

The remainder of the book is organized into five chapters. Chapter 1 begins by presenting general context about the state of electricity access in sub-Saharan Africa, emphasizing how anomalous many countries are relative to the rest of the world. The chapter explains the scale of investment needed to improve electricity access; highlights the fact that large hydroelectric dams are featuring prominently in the energy planning of many countries; and reveals how the electricity provision challenge has created opportunities for new actors to emerge in African energy transitions, in particular, China. Given the state of access and infrastructure, the chapter explains why an analysis of the politics of electricity provision is so important, and proposes a framework for understanding the relationship between politics and energy pathways chosen in countries.

In Chapter 2, the contemporary and historical challenge of electricity provision is examined. The chapter begins by highlighting what is known, and more directly, how little attention electricity has received in English-language academic studies of politics and development in Africa. From this context, the chapter goes on to highlight the regional political history and political economy of electricity and dam construction in East Africa. This chapter shows the currency of historic debates over public versus private delivery of electricity and dam construction, along with debates about electricity for industrial versus individual use.

Chapter 3 examines electricity and energy sector reform in developing countries and Africa specifically, focusing particularly on how energy and electricity fit within the grand period of macroeconomic reform in the 1980s to late 1990s. It highlights what was known and unknown about models of utility and electricity reform during this period and how and why the turn to the private sector became dominant. This is not a revisit of the old stories about structural adjustment and privatization; these issues certainly materialize, but what I emphasize is the disconnect that emerged between the theory of ideal reform versus the reality on the ground.

Chapter 4 examines the contemporary politics of Uganda's electricity sector, including its challenges. Here, the book emphasizes the conflict and tension between sector reform and dam construction and how this conflict produced, in conjunction with dramatic changes in global financial resources, the delays in the execution of the Bujagali dam.

Chapter 5 concludes the book. Here, the epilogue of Uganda's electricity challenges are explained with specific focus on the rise of new 'development partners' like China, which is supporting the construction of three new dams in the country, and turning inward to self-finance its own energy trajectory. This new trajectory, however, is examined in conjunction with a reflection on how politics and governance have evolved in parallel, and what this means for reducing energy poverty and achieving energy justice.

1

Electricity,
Infrastructure
& Dams in Africa

In 2002, only slightly more than 20% of sub-Saharan Africa's entire population had access to electricity, compared to 85% in North Africa, Latin America, East Asia and the Middle East, and 40% in South Asia (Saghir 2005, p. 9). By 2009, the International Energy Agency (IEA) estimated that slightly more than 30% of the population of sub-Saharan Africa had access to electricity (IEA 2011). In many countries, such as Malawi, Uganda, Tanzania, Mozambique, Democratic Republic of Congo and Burkina Faso, less than 15% of populations had access to electricity; only four countries in sub-Saharan Africa of twenty-eight countries documented by the IEA in its 2011 World Energy Outlook – Ghana, Mauritius, Nigeria and South Africa – had access-to-electricity rates greater than 50% (IEA 2011). 'There is a chronic shortage of electricity supply in at least 25 countries in sub-Saharan Africa. At 68,000 megawatts (MW), the entire generation capacity of the 48 countries of sub-Saharan Africa is no more than that of Spain' (ICA 2010). In 2007, the International Energy Agency (IEA) estimated that sub-Saharan Africa (SSA) required $7 billion a year in investment solely for new power generation capacity; if financing for transmission and distribution systems are added, annual investment would need to increase by $30 billion per year (Vedavalli 2007, p. 348).

The gulf between demand for electricity and supply in sub-Saharan Africa has reinvigorated the international community's focus on investments in large-scale electricity infrastructure. In light of the global financial situation that emerged in 2009, the Infrastructure Finance Corporation (IFC), the World Bank, the African Development Bank, and other bilateral agencies established new financing instruments for infrastructure that had a principal aim of facilitating private investment at a time when access to capital was becoming more challenging. Examples of financing instruments included the Infrastructure Consortium for Africa (ICA), the African Development Bank's Emergency Liquidity Facility, the World Bank's Infrastructure Recovery and Assets (INFRA) platform, and the IFC's Infrastructure Crisis Facility. In June 2013, US President Barack Obama also announced $7 billion in financial support and loan guarantees for mobilizing investments in and access to electricity in sub-Saharan Africa. Under 'Power Africa' the US Government aimed to work with the private sector, governments and international institutions to add upwards of 10,000 megawatts of electricity to the sub-continent by 2018. The election of President Donald Trump leaves the future of Power Africa in 2017 uncertain. Amid these investments and announcements, one thing that has been surprising and contentious is the prominence of large hydroelectric dams in these African electricity infrastructure development plans.

Once the physical manifestation of negative, 'high modernist' (Scott 1998) development thinking, large hydroelectric dams have re-emerged as principal engines in national economic and social development strategies.[1] Nyaborongo in Rwanda; Merowe in Sudan; Bui in Ghana; Gibe III and Grand Ethiopian Renaissance in Ethiopia; Grand Inga in the DRC; Kunene in Namibia; the Highlands Water Project in Lesotho; Lom Pangar in Cameroon; Mambilla in Nigeria; Mphanda Nkuwa in Mozambique; Bujagali, Karuma and Isimba in Uganda – these are the names of a small sample of nearly seventy large hydroelectric projects under construction, planning or consideration, which will require enormous capital investments (International Rivers 2010). The turn to large dams as a dominant source of energy generation in sub-Saharan Africa is controversial, but also not surprising for several reasons.

First, the World Bank estimates that of the sites in Africa with current potential to produce hydropower, 93% are 'unexploited' (World Bank 2009). Thus, as some argued years ago, there exists a large volume of 'untapped surplus power' (Ranganathan 1998, p. 3). Given population increase and electricity demand in most African countries, the need for electricity is pressing; electricity is among the highest-priority sectors needing investment in some countries as the absence of reliable supply is known to have undermined investment confidence and hindered the productivity of small, medium-, and large-sized enterprises (Reinikka & Svensson 2001). What is more, there is a severe absence of research on the direct, individual household, and human benefits or impacts of access to electricity in relation to economic activities, education, health and security, and political preferences. Further, research that has been done on the effects of electricity on households is often methodologically weak or unclear as to what is being measured (see Brass et al. 2012). Some research has revealed the indirect benefits from electrification that

---

[1] There is no agreed-upon definition of a 'large dam'. However, a general guide, according to the International Commission on Large Dams (ICOLD), is that any dam that is over 15 metres high is considered large, while a 'major dam' is one more than 150 metres high, has a volume greater than 15 million cubic metres, reservoir storage of more than 25 cubic kilometres, and/or electricity generation of more than 1,000 megawatts (Khagram 2004, footnote 9, p. 217).

were assumed but not demonstrated, such as a more informed citizenry (see World Bank 2010), but this remains rare. Given the World Bank's overarching mission to reduce poverty, the ongoing need for power sector improvement, poor access and the potential for hydroelectric dams to theoretically play a fundamental role in meeting industrial and household electricity needs, a decade ago the Bank reasserted its support for dams by saying that it would 're-engage in high-reward-high-risk hydraulic infrastructure' (World Bank 2004a, p. 3).

Another, second reason that dams are popular investments again is simply due to the fact that they have been used for hundreds of years for flood control, irrigation and to generate power for economic development (see Everard 2013; McCully 2001). Hence, proponents of large dams in sub-Saharan Africa do not hesitate to point to countries like Canada, the United States and Norway to argue that large dams have been critical to economic development (World Bank 2009) and that the consequences of their construction 'must be carefully evaluated against the benefits *by Africans*', not external actors who enjoy access (Meraji O.Y. Msuya, Executive Director, Nile Basin Initiative, interview, 14 January 2003, emphasis added).

Third, global construction of large dams peaked in the 1970s (World Commission on Dams 2000, p. 9), but with the vast majority of dam construction concentrated in developing countries since this time. Khagram (2004, p. 10) explains that declining opportunities for dam construction in industrialized countries, coupled with demand in developing countries and increased access to credit for construction, led private firms to shift attention to developing countries; thus, 'approximately two-thirds of the big dams built in the 1980s and three-quarters under construction in the 1990s were in the third world'.

Owing to the World Bank's central role as a financier of large hydroelectric projects in the 1970s and 1980s, critiques of large dam building in poor countries ran in parallel to critiques of structural adjustment and transparency at the Bank and in Bank-financed projects. One outcome of this era was a proliferation of multinational advocacy organizations pressuring the World Bank to institute more stringent operational policies for projects (including environment and social assessments and resettlement policies and safeguards). A second, related outcome was that the Bank had to begin to heed and formally recognize emerging global norms that promoted and respected indigenous rights, transparency in decision-making, environmental and resettlement policies, and a genuine consideration of alternatives to large dams (Khagram 2004; Leslie 2005; Park 2005; 2009). These global norms began to take solid root following the Bank's controversial involvement in the Indian mega-dam project, Sardar Sarovar, in the mid-1980s (see Khagram 2004; Leslie 2005; Park 2005; 2009), and reached a pinnacle of attention when a global dialogue about the role of large dams in development was organized under the auspices of the World Commission on Dams (WCD 2000).

Rather than deterring future investments in large dams, however, the emergence of these global norms and associated policies designed to alleviate past problems with construction may have, conversely, instilled a new level of confidence in dam construction efforts. Countries and donors supporting or promoting large dam construction cite the existence of these policies and norms to suggest due diligence when funding requires them, even if they are not applied uniformly. Yet, the emergence of these global norms and operational policies also helped produce a shift in the politics and financing of large dams. As the application of these norms became more common, and civil society argued against dams or at minimum the application of safeguards, African governments grew increasingly frustrated with the delays and procedural challenges these policies produced. The result was the rising influence of 'non-Western' sources of financial assistance for dam building. Hence, a fourth reason large dams have become popular is a turn away from the policies and financial requirements of customary developments in the 'West' to 'new' partners, particularly China and firms based in China.

China, large dams and infrastructure financing

China's influence in Africa looms large (Bräutigam 2009; 2015; Michel & Beuret 2009; Taylor 2010). The country has altered the economic and political landscape for how non-Western and Western countries engage with and try to influence African countries, particularly with respect to natural resources (Carmody 2011), but also, increasingly, for many forms of energy provision beyond hydroelectric dams (Power et al. 2016). Along with China, other non-Western or non-traditional development partners are also increasingly prominent in the development of electricity infrastructure in the sub-continent, such as the Arab Fund for Social and Economic Development in Sudan (Verhoeven 2011, p. 135) or the Islamic Development Bank in Uganda (Anonymous, Rural Electrification Agency employee, 15 December 2012). But China's dominance in dam construction in sub-Saharan Africa is unmatched. 'China has emerged as the world's dam superpower' (Verhoeven 2011, p. 123). But in saying this, it is also important to be careful not to generalize about China as one singular entity. As Bräutigam (2015) explains with respect to China's engagement with agriculture in sub-Saharan Africa, fact must be carefully separated from fiction: 'China's engagement in Africa has clearly captured the imagination of many who worry about the impact of a newly rising power on a continent that has seen many foreign invasions ... Getting the details of this engagement right, and avoiding sensationalism, is not easy, but it matters' (2015, p. 10). In Uganda, three distinct Chinese firms are now building large hydroelectric dams. Yet, as I explain below, the participation of Chinese firms in dam building has not gone without domestic controversy or problems. Like Bujagali, the integrity of the bidding process for the dams being built by Chinese firms in Uganda has been questioned. And in 2016, cracks in the foundations of two dams being built raised many concerns about construction quality and construction oversight. China's engagement with large electricity infrastructure in sub-Saharan Africa is not a result of the Chinese government or state firms asserting themselves into African states. Certainly, strategic benefits may accrue to China, but its presence and support for Chinese firms is also a function of a preference and opportunity made available by African states.

China's dominance in infrastructure development generally and dam construction specifically has altered the landscape of interests promoting, financing and building large dams, and is altering the landscape of donor–state relations in African countries. This has led some Western donors to look for ways to remain relevant and influential in the energy pathways and transitions of African countries where China is active in dam building, such as Ethiopia. In Ethiopia, some Western donors chose to focus their support on capacity-building and planning initiatives in electricity as opposed to financing hard infrastructure in order to allow them to continue to have a role in the sector but without being tied to controversial projects (Anonymous, European donor representative, Addis Ababa, 26 November 2010): 'The Chinese are popular with African governments because they build things: infrastructure. Western donors have recently not been keen on roads and ports and are positively allergic to dams... We like the Chinese. When they say they will do something, they do it. 'No consultants, no environmental impact, no delay. You get your road'" (Dowden 2009, Loc 5595 of 6533). Hence, the rise of China's participation in sub-Saharan African dam construction efforts is both pragmatic and strategic.

Domestically, no country has a higher number of large dams in the world than China (WCD 2000, p. 9). In its 2000 final report, *Dams and Development: A New Framework for Decision-making*, the World Commission on Dams reported that China alone had over 22 000 large dams or close to half of the world's total number, while before 1949 it had only 22 large dams (WCD 2000, p. 9). Since the 1950s, the task of building these dams has largely fallen to the state-owned enterprise Sinohydro, which is 'the world's number one hydroelectric company and leading dam-builder in China and across Africa ... Sinohydro's African operations account for 42% of its non-Chinese profits. Besides Sudan, it has built – or is building – dams in 25 other African countries' (Verhoeven 2011, p. 124). In Uganda, mirroring the controversy and role of the President in the Bujagali dam, Sinohydro Corporation Ltd was picked to build the next dam on the Nile, the Karuma dam, despite much controversy internally over whether the contract was awarded properly (Wakabi 2013). Subsequently, the China Water & Electricity Corporation (CWEC) won the contract to build the Isimba dam, and the

Gezhouba Group won the contract to build the Ayago dam – all large dams. CWEC is a subsidiary of the state-owned company, China Three Gorges Corporation, the company established in 1993 to build the Three Gorges Dam. Each of these three corporations is directly or indirectly state-owned, and all projects are being financed with loans from the Export-Import Bank of China and the Government of Uganda. Hence, Uganda is deeply committed to Chinese investment and participation in its future electricity generation system.

Despite high media and academic attention to China's engagement in Africa, there is a very long history of foreign governments leaving a lasting mark on the rivers and hydroelectric landscape in the sub-continent. At the turn of the last century, Great Britain governed territories containing more than half of the world's big dams (Khagram 2004, p. 5): 'British colonialists were the most ardent dam builders outside Europe and North America in the late nineteenth and early twentieth centuries, leaving their mark most firmly on the basins of the Indus, Ganges and Nile' (McCully 2001, p. 18). France was equally active in its North African colonies, with control over water serving as a mechanism to control the colonies (Pritchard 2012).[2] Given this history, what makes China's presence different?

China's role in dam construction and natural resource extraction is not a new form of hydro-imperialism, for the country has little interest in influencing the domestic political affairs of African governments (Carmody 2011). China's influence may well be strategic and opportunistic but it also serves as an alternative and preferred 'development partner'. Its influence is of longer standing than most acknowledge, but also a result of a shifting set of relationships between Western donors and African governments when it comes to dam and infrastructure construction. Because China does not abide by the same operational policies that the World Bank and Western bilateral donors follow when undertaking or financing infrastructure projects, African states have viewed it as a desirable 'partner' – the speed of project review and execution can be faster. Indeed, as will be noted later, the experience with multilateral donor policies in some East African countries in the mid-2000s led governments to purposefully look for alternative financing arrangements. The potential benefit of working with China is further heightened in African countries that are transitional democracies, that have weak concentrations of civil society organizations, or that have simply grown frustrated with the time it takes to implement projects following global norms and policies.

[2] Richard Dowden (2009) writes that when the British took over Sudan in 1899 this had little to do with Sudan 'and everything to with India, the "jewel of the crown" of the empire' (Loc 1825 of 6533). The British feared that the French might be able to stop or divert the Nile, thus eliminating easy access to India via the Suez Canal. Thus, 'the British became convinced they must control the Nile from mouth to source in case another European power took it and threatened the route to India' (Loc 1831 of 6533).

China's emergence as a prominent player in African dam construction thus is a result of three complementary dynamics: Chinese advocacy and entrepreneurialism; African government frustration with and/or resistance to the processes of infrastructure construction and financing required by Western lenders, particularly the World Bank; and the capability and willingness of African governments to govern their territories and establish conditions that promote or deter Chinese investment (see Mohan & Power 2008; Mohan & Lampert 2012). The presence of China and other non-Western lenders in African electricity sectors therefore has important instrumental and political outcomes.

Sub-Saharan African countries now have more options for project financing and execution than previously and can look away from traditional Western lenders – multilateral and bilateral agencies and Western Export Development Corporations – to support and execute large infrastructure projects. African governments are not passive; the presence of the Chinese state and Chinese firms requires an understanding of the African institutions that are doing business and turning to Chinese firms (see Mohan & Power 2008), hence the changing nature of African governments in development choices.

The presence and role of private firms in dam construction is decades old, but historically, if private firms were to independently lead dam construction initiatives or were to partner in dam construction, they would have to have project financing underwritten by bilateral or multilateral development finance institutions such as the Multilateral Investment Guarantee Agency (MIGA) to cover economic and political risks. China's financial resources and experience in dam construction thus offers African governments greater choice in assessing *who* will be their electricity 'development partners', but also *how* electricity expansion will take place. It is the 'how' of electricity provision and expansion and the conflict and political change that ensues in that process that is a central concern in this book.

Why study the politics of electricity and infrastructure and how?

The politics and conflict of dams and infrastructure development have been of interest to researchers for decades. In his classic book, *Development Projects Observed*, Albert O. Hirschman (1967) examined the role of the state in project implementation and introduced the alluring notion of the 'Hiding Hand' – an invisible hand that emerges in project development to conceal project difficulties until implementation is well under way. Hirschman observed that project planners often underestimate the costs of projects and, when confronted by implementation difficulties, press harder for the project to be completed: an underestimate of project difficulties is required 'so that perfectly feasible and productive projects will actually be undertaken' (Hirschman 1995 [1967], p. 17). The veracity of the 'Hiding Hand' has since been challenged (Flyvberg 2014), but the basic premise of the argument – that proponents of megaprojects proceed while underestimating the difficult of the undertaking – remains. Other work in the same time period acknowledged the political importance of hydroelectric dams:

A hydroelectric project is fine political capital. The politician looking for a good public works project is much more likely to select power if it is hydro. The hydro complex has drama and style, and there is an air of extravagance in its hugeness and grace which is awesome in a country trying to mobilize scarce resources for development. Though hydro supplies a basic necessity, it creates the aura of a country which no longer has to scrimp and save, but can spend with largesse. Its hugeness and its taming of a wild river bespeak a technological victory, and it imparts dignity to the people and the country who conceived it. (Tendler 1965, pp. 250–1)

Even at a time when Uganda was ruled by notorious leader, Idi Amin – a man who publicly lashed out at the British – the Owen Falls Dam, built by the United Kingdom during the colonial period, appeared prominently on Uganda's currency (see Figure 1).

Later work continued to acknowledge the political attraction of larger hydro projects and mega-projects generally. Scott referred to the mythical appeal hydroelectric projects hold for leaders (1998, p. 166), while Flyvbjerg (2014, p. 8) notes the 'political sublime' of mega-projects: 'the rapture politicians get from building monuments to themselves and for their causes'. The early research of Hirschman encouraged an examination of the process of project implementation. But it was also later criticized by some scholars suggesting that there seemed a general acceptance of the projects that the 'development apparatus' promoted and not enough critical reflection on the logic and mode of the 'apparatus' itself: the 'development apparatus' does not make 'its effects felt only through documents and reports, but also through policy, programs, and most characteristically, "projects"' (Ferguson 2005 [1994], p. 74). For some years then, an examination of the relationship between politics and technological or technocratic solutions to 'development problems' has been deemed important, but surprisingly under-examined, particularly in relation to hydroelectric dams and electricity in Africa (one early exception is McDonald 2009). Indeed, in a 1980 article on the Aswan High Dam the authors wrote that 'we know far too little about the linkages between technical and political aspects of the decision-making process, the problems of selecting and assessing policy options that have large technological components, or the appropriateness of various theories of decision making to technical controversy' (Rycroft & Szyliowicz 1980, p. 36). Hence, despite more than three decades of political, economic and social sector reform in sub-Saharan

Figure 1 Uganda currency circa 1979, showing Owen Falls Dam

Africa, there still remain few examinations of the politics and conflict surrounding reform processes, project implementation and transformations associated with energy (see Keeley & Scoones 2003; Brock, McGee & Gaventa 2004). Why, for example, do governments choose to build hydroelectric dams over other electricity generation options? How do dams factor into the mix of other solutions to energy problems? Do 'big' problems require 'big' technical solutions?[3]

The limited examination of the politics of mega-projects (Flyvbjerg 2003) is significant given that dam and electricity expansion projects

---

[3] I am indebted to the late University of Toronto professor, Dr Rodney White, mentor, colleague and friend, for asking me this question early on in my research.

are taking place in a much different social and political context than in the past: re-regulation, unbundling of government monopolies and privatization commonly accompanied dam construction and electricity expansion efforts in the late 1990s (Verhoeven 2015 and Isaacman & Isaacman 2013 are important recent contributions to the study of African dams). Of course, mega-projects do not occur in a vacuum. They are often taking place at the same time as, or must be combined with, other institutional and legal changes. Hence, major infrastructure projects are usually 'mega-undertakings' – undertakings that combine hard infrastructure investments with widespread policy, programme and institutional reform. These kinds of 'second-generation reforms' contrast markedly with first-generation macroeconomic reforms like currency devaluation that could be done by a 'stroke of the pen' and did not encourage, require or solicit a high degree of public scrutiny (Brink-erhoff & Crosby 2002, p. 22).

Dam construction has always solicited controversy and critical consideration owing to human displacement, the presence and promi-nence of international and domestic civil society organizations, and, in recent decades, greater access to information and ease of communica-tion. But since the 1990s, and particularly when dam construction has been combined with major public sector and economic reforms, as well as a greater willingness on the part of domestic civil society to engage in debate with governments or donors, controversy over dams has emerged again. The result is increased conflict domestically, but in a manner that remains unorthodox in many countries: state–society and state–international conflict now often materializes in formal institutional and legal forums, such as in Parliament and Parliamentary committees, courts, and international forums like the World Bank's Inspection Panel procedures. In the case of Uganda, when domestic and international NGOs used these forums to challenge the Bujagali dam, they were criti-cized for their self-interest and blamed as the reason for Bujagali failing to be built in a timely manner (Mallaby 2004) – a view the President of Uganda wholly agreed with, calling NGOs 'economic saboteurs' and 'enemies of the state'.

When governments are frustrated with project implementation processes that are time-consuming and that reveal the hidden costs or realities of project activities, then the potential for conflict to emerge in the implementation of projects and reforms escalates – processes that on paper are thought to be rational and linear and easily replicable (Hirschman 1967) but in reality of course are context-dependent. In Uganda, donors and the World Bank, along with President Museveni, assumed and argued publicly that electricity reform and the dam could evolve in a short period of time, in a linear manner, and would be largely technical in nature, despite global evidence that this is not usually possible (see Gaventa 2004; Grindle 2004; McGee 2004). While some bilateral donor representatives deeply familiar with the energy reform process in Uganda in the early 2000s conceded that connecting sector reform and dam construction was highly risky, the process continued. But in retrospect, and as will be explained later, more than one World Bank official deeply familiar with Uganda's energy reform experience in the early 2000s admitted in confidence years later that the construction of a large dam and the reform of a sector at the same time in Uganda was too risky and progressed too quickly. Indeed, Uganda's experience with electricity reform in the early 2000s is a near-ideal 'critical case' of Easterly's (2013) argument about what happens when technical goals and expertise clash with a powerful national leader who dismisses opposition and debate as 'anti-development': the desires and goals of the individuals and poor are overwhelmed and lost in the race to implement projects.

Some multilateral development agencies have openly acknowledged that 'it's time to look beyond the specific content of policies to the *critical processes* that shape these policies, carry them forward from idea to implementation and sustain them over time' (Inter-American Development Bank 2005, p. 1, emphasis added). In the mid-2000s, at the height of the privatization and energy sector reform period in sub-Saharan Africa, the World Bank also suggested many procedural and process-related conditions deemed necessary for successful reform (World Bank 2004b). Despite this, one of the central issues often overlooked in electricity reform is that these conditions – these 'critical processes' – have remained subservient to technical and regulatory goals with the political context and shifting political character of countries not taken into consideration. The politics and process of state–society interactions during reform and project implementation are treated as secondary concerns to technical matters, and, in turn, are put aside in the rush to execute a reform. In doing so, the 'indirect effects' of the process leading to the execution of the project or reform are not well considered; as Albert Hirschman noted four decades ago, ignoring the indirect effects of a project implementation process may inflict penalties that are anything but nebulous (Hirschman 1995 [1967], p. 163). Indeed, as one former, senior member of the Ugandan Ministry of Energy remarked in 2008 when I asked him about the World Bank's influence and reform guidance, he said: 'We swallowed the gospel and have moved on' (Interview, senior government official, Ministry of Energy, 23 June 2008). In Uganda, 'moving on' meant looking to new 'development partners', turning away from the World Bank's energy sector advice, and continued public tension over the dismal quality and access to electricity in the country. Given the significant direct and indirect effects of the energy sector reform process, it is imperative to understand how the national political context explains which energy '... pathways get supported and legitimized, and which are ignored and fail to gain traction' (Scoones, Newell & Leach 2015b, p. 7).

Conceptual framework and approach

This book examines the politics of electricity sector reform and dam construction in sub-Saharan by focusing on the processes through which local, national and international interests and actors converge in decision-making and deliberation intersect. National governments have the unenviable task of trying to mediate successful policy interventions from both international expert bodies and marginalized groups (Forsyth et al. 1998, p. 38). This situation is further complicated by the evolving and maturing bureaucratic systems in African countries, which are trying, and being forced, to learn (at an historically unprecedented pace) how to respond to a more vigilant, informed and globally connected citizenry, while also trying to administer reforms. In the late 1990s and early 2000s, many national governments in sub-Saharan Africa were highly influenced by multilateral development agencies, and, for some, international donors were better understood as political actors deeply embedded in the African state, if not part of it (see Harrison 2001; 2004a; 2005). Harrison noted, for example, that in the early 2000s, politics in African countries had been dominated by 'donor dependency' (2001, p. 660): 'Rather than conceptualizing donor power as a strong external force on the state,' Harrison suggested that it was 'more useful to conceive of donors as *part of the state itself*. This [was] not just because so much of the budgeting process [was] contingent on the receipt of donor finance, but also because of the way programmes and even specific policies [were] designed and executed' (Harrison 2001, p. 669, emphasis in original).

The evidence from electricity sector reform in Uganda in the 1990s and early 2000s supports this general contention. In Uganda, the World Bank led and promoted the country's reform agenda. Other bilateral agencies provided support for capacity-building in affected ministries, but the overall rationale and structure of the 'mega-undertaking' was World Bank-led. As is later detailed in the book, in Uganda, this is not a controversial statement: evidence for this position comes from interviews with the bilateral agencies, the private firm originally building the Bujagali dam, and senior government officials in Uganda. Given this, to understand the politics and process of electricity transformation requires researchers to understand politics in a qualitatively different way (Harrison 2001, p. 661). There is a need to conceptualize the past and current character of relations between state and non-state actors in a manner that captures the reality of how multiple interests from the international through to the local level interact and, as Hirschman noted, 'the indirect effects' of those interactions. As earlier noted, recent work (Baker, Newell & Phillips 2014; Newell & Bulkeley 2016; Newell & Mulvaney 2013; Newell & Phillips 2016; Power et al. 2016; Scoones, Leach & Newell 2015a) argues convincingly that a political economy approach to energy transitions provides an appropriate framework for understanding relations of power between actors along with the structures governing outcomes.

I endorse this view, but do not apply it as an organizing framework for this study. While I am centrally interested in the power relations between interests that shape technology and reform choice, I am principally concerned with the process of interaction – the character of the relations between actors and how they deliberate (or do not deliberate) various energy pathways. Given this, I argue that the notion of 'governance' offers 'a qualitatively different way' of understanding politics and the processes of energy reform. The notion of governance, and the character of 'energy governance', serves as a valuable overarching analytical framework for understanding how reform takes place and why and how these processes shape politics and policy outcomes.

My use of 'governance' is not derived from a central concern with improved public management or corruption – a view often associated with new public management and the World Bank in years past. My focus is on understanding and characterizing how state and non-state interests interact over energy and electricity issues in order to understand how these relations influence politics and outcomes. Using 'governance' to frame the study of electricity transformations builds on more than a decade of influential (and divergent) uses of the term in development and African studies, along with conceptual discussions of the concept in political science and political theory (see Ericson & Stehr 2000; Hyden & Bratton 1992; Hyden et al. 2000; Kjær 2004b; McCarney et al. 1995; McCarney 1996; McCarney 2000; McCarney & Stren 2003; National Research Council 2003; Olowu & Sako 2002; Pierre 2000; Stren & Polèse 2000). One of several reasons that the concept is attractive for the study of energy and political transformations is that it emphasizes that the state is not the locus of decision-making authority (Lofchie 1989 in McCarney et al. 1995, p. 94), and embraces the argument by energy scholars that outcomes relating to energy are a result of a multidimensional network of interactions contingent on resource endowments and settings (Day & Walker 2013). This observation is particularly important for studies of African countries, where a state's ability to provide services is frequently challenged by a lack of capacity and resources, and because non-state actors have an indelible impact on policy decisions and a role in providing services.

Following McCarney, Halfani and Rodriguez (1995), the notion of 'governance' in this study is defined as the 'character of relations between state and non-state actors'. By first understanding the character of relations between state and non-state interests and the conditions that shape those interactions, researchers can then proceed to understand how policy and programme choices materialize from those interactions to shape policy outcomes (see Hyden & Court 2002) and for the energy sector, the pathways chosen. Governance then, is used as an

Figure 2 Analytical framework for analysing energy pathways and policy choices

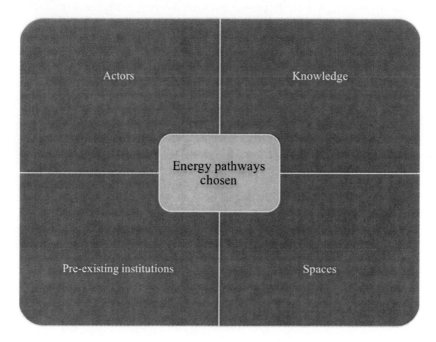

objective empirical framework to describe relationships generally can be broken down and applied to specific policy or decision processes in order to understand the influence of different actors in the process of decision-making.

Building on categories developed by researchers at the Institute for Development Studies (IDS) at Sussex University, I use four factors to help understand the interaction between state and non-state actors in the process of choosing energy pathways and policy choices: (1) the actors included in and excluded from decision-making; (2) the structure or space of decision-making, considering, for example, such issues as the forums for public debate and opportunities to contribute to these forums; (3) the knowledge included and excluded from decision-making; and (4) the role of pre-existing institutions in defining the formal and informal 'rules' that characterize the sector and rules of decision-making. These categories build on the observation that policy is a dynamic process, which is influenced by actors from the international through to the local level, the knowledge they carry, and the spaces in which they interact (McGee 2004, p. 8). Politics, policy choices and pathways do not exist in a national vacuum (McGee 2004, p. 22), and are contingent on history and global political economy. Recognizing this, Figure 2 presents the relationship between governance and the energy pathways and policy choices chosen.

Using this framework permits an observer to not only identify which actors were influential but how they exercised that influence. For example, it is not a surprise that the World Bank was a dominant policy actor in Uganda at the height of electricity sector reforms. But using this framework provides guidance for understanding more deeply the various ways that the World Bank decision to endorse a 'big bang' approach to reform materialized procedurally. How did it convince the national government to undertake such an ambitious agenda? Did the national government have a choice? What opportunities for broader deliberation and debate were provided? If other bilateral agencies believed the World Bank's agenda was too risky, why were they weak or powerless to alter that agenda? Hence, this 'governance approach' aims to open the 'black box' of state decision-making in Africa, and to better illuminate the multilevel politics of policy and energy pathway choice in the sub-continent.

We now turn to the history of electricity provision in sub-Saharan Africa generally and East Africa specifically, revealing that several historic debates and conflicts over electricity provision have persisted for decades.

2

The Politics of Provision

A History of Debate & Reform

'How should a developing country government, concerned with tackling poverty amongst its citizens, think about its role in the energy sector?' (Bond, in World Bank & ESMAP 2000, p. vii). This simple question has proven vexing for countries in sub-Saharan Africa. Energy is a particularly vexing policy area because of its connection to so many other sectors and because of the varied and mixed ways that the benefits of energy sector reform are communicated and can be communicated to citizens. What is, for example, the primary reason for introducing private sector providers? For increasing electricity tariffs? For building a hydroelectric dam in lieu of investing in decentralized renewable projects? Hence, the questions are not only about how a developing country government should think about its role in the energy sector but also what the implications are of the energy pathways or investments chosen. How, for example, does a government address the inequity in access to modern energy services? How are questions about the unequal distribution of electricity services communicated? How are policy and reform decisions framed, assigned meaning and communicated?

One way that these questions have been examined is through the study of 'narratives' or, in this case, energy narratives. As Roe (1991) notably highlighted, a narrative is a story with a beginning, middle and end. What makes a narrative powerful is not its veracity, or the evidence to support it, but the strength or power of the narrative in driving actions and convincing others of the merits of the action. A narrative does not need to be true; but if the narrative is powerful and there are no other narratives that can compete and usurp dominant narratives, then the dominant narrative persists. For environmental issues in Africa generally, dominant development narratives have historically produced inaccurate and sometimes destructive policy interventions (see Leach & Mearns 1996). Hence, it can be asked: what is the dominant energy narrative being promoted by governments and international organizations? How are policy reforms or investments being framed and assigned meaning?

Are historical, social, political and economic conditions or contexts in African countries amenable to the energy narratives being promoted?

Shortly after Uganda's state electricity utility was unbundled, the government produced a national energy policy. The 2002 policy acknowledged the link between energy and other sectors, sub-sectors, the economy, and regional and international influences. It also articulated the relationship between the provision of energy and a reduction in poverty. But acknowledging this relationship is very different than implementing reforms and policy preferences that explicitly, purposefully and directly address widespread energy poverty and energy justice? Will reforms aim to improve access for all? Will the quality of service be equitable? What about the timeline for access?

In recent years, the World Bank and the Energy Sector Management Assistance Programme (ESMAP)[1] have examined the notion of 'energy access' more closely and developed a multi-tier framework for measuring energy access which accounts for electricity quality and the human impacts of various energy sources (ESMAP 2015). This work is timely but it does not alleviate the tensions that exist between governments and international agencies over the simple but profound question about who energy reform is for. International organizations have been concerned with this question for years. Indeed, in a 2000 publication, ESMAP asked how pro-growth, pro-efficiency reforms should be weighed against those of direct interventions aimed at improving the poor's access to modern energy for consumption and productive uses (World Bank & ESMAP 2000, p. 2). Should electricity sector reforms directly attempt to improve energy services for the poor, or do the poor have to wait until the benefits of improved electricity services for industrialization produce indirect benefits like job creation and increased income, which might eventually translate into the capability to pay for electricity services?

In the early 2000s, officials in Uganda's Ministry of Energy and Minerals Development (MEMD) were also struggling with these questions. They acknowledged that all reforms had to be looked at in terms of the country's primary objective of poverty elimination. What was contested, however, was how poverty alleviation through energy reform would be realized – at what pace would benefits accrue to citizens and at what cost to the government? Ugandan Ministry of Energy officials expressed that providing the necessary input for industrialization – modern energy – was the first priority in reforms, with electricity to homes, domestic users and the poor a secondary priority (Interview, Moses Murengezi, Ministry of Energy, 13 January 2003). Uganda's former Commissioner of Energy, Godfrey Turyahikayo (Executive Director of the Rural Electrification Agency in 2017) confirmed this when he

[1] ESMAP was established in 1983 under joint sponsorship of the World Bank and the UNDP. Today, ESMAP provides technical assistance and policy advice on sustainable energy development to governments in developing countries and economies in transition. For more information, visit www.esmap.org.

explained that individuals were a 'second priority' in sector reforms in the early 2000s, and that the purpose of reforms were to drive the economy with the spin-off effects of industrialization producing jobs that would lead to improved individual well-being; this, he acknowledged, was 'one of the hard facts of development – you have to ignore individuals and arguments for individual power' (Interview, Godfrey Turyahikayo, 5 June 2002). Yet, on more than one occasion, I bore witness to government and donor representatives disagreeing about whether connecting households to the electricity grid should be prioritized. Owing to the complexity and cost of connections, government representatives argued for expanding 'access', understood as proximity to an electricity supply as sufficient, whereas donors were arguing for direct household connections. Government officials were not convinced that there was any benefit to providing electricity to individuals who may not be able to afford it, or who only had a light bulb or radio.

This tension revealed an important conundrum: if the central purpose of electricity reforms in the eyes of government is industrialization, yet multilateral agency research and policy documents recommend household connections with improved quality, what materializes in practice on the ground? How are these tensions over the path of electricity reform reconciled? And what knowledge and historic evidence is used to support particular reform pathways?

When human development is understood as a process of providing citizens with increased opportunities to make choices in their daily lives in order to gain 'social power' – the power needed to engage in economic, social and political activities (Friedmann 1992) – then the case for providing electricity to households is strengthened. One of the eight bases of social power, according to Friedmann, is 'surplus time over subsistence requirements'. Given that access to electricity can provide citizens, particularly women and children, with the opportunity to gain 'surplus time' over activities like firewood collection, access to electricity seems to be a strong potential contributor to social power. Understanding 'development' in this way – as a function of the freedom and opportunity an individual or household has to make decisions over their life (see Drèze & Sen 1995; Sen 1999) – makes the case for electricity to households stronger. With respect to energy justice, this perspective means that energy reforms ought to 'maximize welfare … and that every person has the right to a "social minimum" of energy or electricity' so that the individuals can equally realize their capabilities (Sovacool & Dworkin 2014, p. 245). In practice, governments in Africa with limited financial resources baulk at this argument. And for advocates of this conception of energy justice, there is still very little strong empirical research that shows the correlation between electricity access and improved human welfare. This is assumed but not well documented, with many methodological challenges when trying to show the effect of electricity on citizen behaviour and welfare.

For example, research reveals that regular access to information and knowledge gained via communication technologies powered by modern energy systems – radios and television – is directly related to citizen knowledge about domestic and global political issues and helps in the development of a 'critical citizenry' – a citizenry that adheres to democratic values and has a healthy, critical scepticism of leaders and institutions (see Norris 2001; Mattes & Shenga 2007; Moehler 2008; Moehler & Singh, 2011). Using a series of multiple regression models and controlling for level of education, research in Mozambique demonstrated that 'watching news programs and listening to them on the radio (but, notably, not by reading newspapers) ... makes an important, independent contribution' to access to political information and the development of a critical citizenry (Mattes & Shenga 2007, p. 28).

Thus, evidence exists that electricity does provide 'connective power' to households (Jacobsen 2007) – a long-held but rarely tested hypothesis. That is, through regular access to information and knowledge via technologies that are connected to a steady supply of electricity, as opposed to costly battery-operated devices, citizen knowledge and information about political institutions increases and becomes more critical. Despite these significant advances, other public good benefits from household access to electricity such as improved social cohesion and political stability (Tendler 1979), improved security from lighting, or increased participation or engagement in public affairs due to time savings or greater knowledge continue to be assumed but not evaluated systematically (World Bank 2008, p. 45).

There is also some evidence that democratization leads to increases in the distribution of electricity to ordinary citizens (Brown & Mobarak 2009, p. 194). However, the question of whether increases in electricity access (and use) lead to higher levels of participation in civic and political life remains largely unanswered. Limited qualitative research suggests 'yes', but there is a need to understand the connective power of electricity more rigorously (Jacobsen 2007, p. 156), particularly in sub-Saharan Africa, where one of the acute scarcities, especially in rural areas, is lack of public information (Bratton, Mattes & Gyimah-Boadi 2005, p. 347). With regular access to information and knowledge through mechanisms powered by electricity, increasing evidence and assumptions suggest that citizens will gain more knowledge about democracy and participate more actively and critically in political institutions (Bratton, Mattes & Gyimah-Boadi 2005, p. 348). While observational data and qualitative interviews can provide valuable context about the role of electricity, there remains a need for more systematic research on the effect of electricity in order to understand the social and political significance of electricity in a broader national context (Jacobsen 2007, p. 156).

The other major methodological dilemma is what is meant by electricity or modern energy – what form should it take and what quantity and quality is justifiable? ESMAP's multi-tier framework for measuring energy access provides a way to operationalize variation in access, but how is this variation considered in decision-making? Again, there remains very little research on this subject, and research that has been done, particularly in relation to the provision of non-grid based electricity, has been poor. Brass et al. (2012) reviewed distributed generation projects in the developing world and found several critical problems with existing research. Research, for example, has provided little indication of what constitutes success or failure in projects; whether project implementation and outcomes are treated differently in assessments; and what methods of assessment are used (ibid.). Research papers also often fail to articulate a clear research question, hypotheses, clear research designs, and transparent methodology or data. Given this serious methodological problem, it follows that researchers are left to wonder what evidence or rationale countries and development agencies use when choosing different energy pathways.

In this chapter, debates about electricity access are put in historical context. The chapter focuses specifically on how electricity access, provision and expansion have been conceived or framed in the late pre-colonial, colonial and early post-colonial era in East Africa. The chapter reveals that debates about expansion and access, and the role of private interests versus government are not at all new. Indeed, as the book will reveal later when examining contemporary reform, there is an element of 'déjà vu all over again'.

Early electricity in East Africa: Government-or private-led?

In December 1905, Winston Churchill, then 31, was appointed to his first ministerial post as Parliamentary Under-Secretary of State for Colonies. A short time later, in 1907, Churchill toured East Africa during the parliament's autumn recess. Writing in his 1908 travelogue *My African Journey*, and in reference to his July 1907 visit to Uganda and the Nile, Churchill debated the role of public and private interests in the development of Uganda's hydroelectric potential:

> As one watches the surging waters of the Ripon Falls and endeavours to compute the mighty energies now running to waste, but all within the reach of modern science, the problem of Uganda rises in a new form on the mind. *All this waterpower belongs to the State. Ought it ever to be surrendered to private persons? How long, on the other hand, is a Government, if not prepared to act itself, entitled to bar the way to others?* This question is raised in a multitude of diverse forms in almost all the great dependencies of the Crown. But in Uganda the arguments for the State ownership and employment of the natural

resources of the country seem to present themselves in the strongest and most formidable array. Uganda is a native State. It must not be compared with any of those colonies where there is a white population already established, nor again with those inhabited by tribes of nomadic barbarians. It finds its counterparts among the great native States of India, where Imperial authority is exercised in the name and often through the agency of a native prince and his own officers.

... In such circumstances there cannot be much opening for the push and drive of ordinary commercial enterprise. The hustling business man – admirably suited to the rough-and-tumble of competitive production in Europe or America – becomes an incongruous and even a dangerous figure when introduced into the smooth and leisurely development of a native State. The Baganda will not be benefited either morally or materially by contact with modern money-making or modern money-makers. When a man is working only for the profits of his company and is judged by the financial results alone, he does not often under the sun of Central Africa acquire the best method of dealing with natives; and all sorts of difficulties and troubles will follow any sudden incursion of business enterprise in the forests and gardens of Uganda. *And even if the country is more rapidly developed by these agencies, the profits will not go to the Government and people of Uganda, to be used in fostering new industries, but to divers persons across the sea, who have no concern, other than purely commercial, in its fortunes. This is not to advocate the arbitrary exclusion of private capital and enterprise from Uganda. Carefully directed and narrowly controlled opportunities for their activities will no doubt occur. But the natural resources of the country should, as far as possible, be developed by Government itself, even though that may involve the assumptions of many new functions ... Nowhere are the powers of the Government to regulate and direct the activities of the people more overwhelming or more comprehensive.* (Churchill 1989 (1908), pp. 75–7, emphasis added)

Churchill believed that the Nile represented an untapped opportunity for industrialization in Uganda, and envisioned 'the gorge of the Nile being one day crowded with factories and industries', given that he saw there being 'power enough to gin all the cotton and saw all the wood in Uganda' (1989, p. 74). 'It would be perfectly easy,' he said, 'to harness the whole river and let the Nile begin its long and beneficent journey to the sea by leaping through a turbine' (1989, p. 75). These remarks represented a turning point in the history of Britain's aspirations in Uganda as well as for the entire East African Protectorate. Further, Churchill's conviction about the role of the state over private interests is noteworthy.

Tanzania (Tanganyika) and Kenya had both established electricity companies prior to Uganda. According to the Kenya Power and Lighting Company (KPLC), electricity was first established in East Africa by the

Sultan of Zanzibar, Seyyeid Bargash, in 1875. Bargash acquired a generator to light his palace and the nearby streets of Stone Town. In 1908, a wealthy merchant from Mombasa, Harrali Esmailjee Jevanjee, acquired the generator and then transferred it to the coastal city for use by the Mombasa Electric Power and Lighting Company. In the same year, an engineer, Clement Hertzel, was granted exclusive rights to supply electricity to the then district and town of Nairobi, which led to the formation of the Nairobi Power and Lighting Syndicate. By 1922, the two utilities in Nairobi and Mombasa were merged under a new company incorporated as the East African Power & Lighting Company (EAP&L).

By 1932, the EAP&L had also acquired a controlling interest in the Tanganyika Electricity Supply Company Ltd (TANESCO), which, according to Charles Hayes, author of *Stima: An Informal History of the East African Power & Lighting Company*, was consistent with the company's plan to move quickly over its Kenyan borders to make power supply truly East African (1983, p. 315). At the same time, Hayes writes that the Company was also investigating licences for generation and distribution in Uganda, particularly for Ripon Falls, Jinja (site of the Nalubaale dam) and Kampala. A lack of financial resources hampered the execution of these early goals, however. As a result, in the early 1920s and 1930s, individual and small private company efforts to develop electricity or acquire generation rights unfolded spottily near towns and trading centres in Tanganyika and Kenya. Efforts to create more integrated and stable electricity generation and supply networks would continue well past independence for both countries, with limited financial resources and hydroelectric generation potential stirring ongoing debates over the merits of large-scale versus small, incremental investments in generation and distribution systems. As a result, given the apparent vastness of Uganda's hydro resources, during this period the EAP&L continued to look to Uganda as a potential secure source of electricity.

The history of the EAP&L's formal presence in Uganda along with its relationship to its predecessor, the British East Africa Company, is not clear, but Hayes writes that in 1904 the EAP&L had articulated in a prospectus the possibility of erecting a generating station at Ripon Falls. Three years later, Churchill would reemphasize this possibility but the Uganda Secretariat apparently rebuked EAP&L's initial interest in the early 1900s (Hayes 1983, p. 329). While the Secretariat is thought to have considered the early proposal it was not ready to grant a concession to the company at the time (Hayes 1983, p. 330). This early hesitation to grant licences to the private company indicates one of the first points when questions were being formally raised over the viability and role of private versus government-led development of an electricity supply in Uganda. Nonetheless, the EAP&L would have to wait until the mid-1930s for its proposal to be formally reconsidered.

By 1936, Harold Odam, head of the EAP&L, secured an interview with the Ugandan Governor, Philip Mitchell, to discuss generation and distribution opportunities in the Ugandan Protectorate. At the meeting, Odam proposed to construct three thermal generating stations in key, large southern Ugandan cities – Jinja, Kampala and Entebbe – in order to immediately service each of the areas and to maintain an option to later develop the Nile's hydroelectric potential near Jinja. As Hayes reports, Mitchell thought Odam's initial proposal was absurd given that the development of one hydropower source on the Nile could satisfy all electricity needs in lieu of the many small, decentralized generation sources he was proposing. What is more, the hydropower source Mitchell had in mind was none other than a dam at Bujagali Falls.

Odam's reluctance to embark on the immediate construction of a large hydroelectric facility lay in a concern that was later debated in Uganda in the early 2000s – the EAP&L was concerned about the potential market for electricity in Uganda, particularly given the absence of industry that would be the largest consumer (Hayes 1983, p. 330). Odam's concerns were also consistent with the findings from a 1935 survey of the hydroelectric development potential of the Nile conducted by the future Chairman of the Uganda Electricity Board, C. R. Westlake. Westlake argued that, while technically feasible, a large hydroelectricity project on the Nile 'would not pay, as electricity consumption both actual and potential, was too low' (Wilson 1967, p. 2).

Despite Governor Mitchell's ambitions for a large dam he conceded to the EAP&L's proposal and granted the company licences for thermal generation and distribution in each of the large southern cities of Uganda. Commercial service became available in Kampala and Entebbe by 1938, and shortly after in Jinja. EAP&L's monopoly would last less than ten years, however, during which time Britain's desire for a much grander hydroelectric project only grew. Three debates stand out in this early colonial period: (1) whether large generation versus small distributed generation sources were most appropriate; (2) whether electricity generation sources should be built ahead of demand; and (3) whether infrastructure development should be led by government or private interests.

Electricity in Uganda: Colonial and early post-colonial legacies

While Westlake was undertaking his survey of the Nile's hydroelectric potential in Uganda, another, equally influential assessment of the economic development opportunities in the country was under way. In 1945 Sir John Hall became Governor of Uganda. Hall took up his position at a time of tremendous change in Uganda and globally following the end of the Second World War. His goal was to see Uganda develop a vibrant export sector based on agriculture with as much industry as possible (Wilson 1967, p. 1).[2] This vision was formally articulated in the 1946 *Uganda Development Plan* produced by Dr E. B. Worthington. Worthington's report did articulate the need for electricity, but it did not factor significantly. The explanation for this lack of attention would come a year later with C. R. Westlake's publication of the *Uganda Electricity Survey*, which Wilson notes 'was as much a marketing survey as a technical report' (Wilson 1967, p. 2).

> Westlake's recommendations, very much in line with the beliefs of the Uganda Governor, Sir John Hall, were presented to and adopted by Uganda's Legislative Council in July 1947. The prospect was awesome – nothing less than a £22 million plan for harnessing the Nile at its outflow from Lake Victoria Nyanza and the creation of a new authority, the Uganda Electricity Board. (Hayes 1983, p. 331)

The government wasted little time in implementing its vision. On 18 January 1948, the Uganda Electricity Board (UEB) was formally created as a quasi-independent vertically integrated monopoly to generate, transmit, distribute and supply electricity within Uganda, with a vision to supply the wider East African region. Westlake was appointed as UEB's first Chairman, which quickly took over the private company, EAP&L. Taking over the EAP&L's generation and distribution activities in Uganda was facilitated by the fact that the company was encountering serious service delivery problems, with coffee and curing companies writing to the Director of Public Works and Chief Secretary of Agriculture in 1945 to complain about stoppages in production due to power outages (see Figures 3 and 4).

Together, Hall and Westlake had produced a vision for the future development of Uganda that rested on a large-scale, government-led plan to develop its hydroelectric resources: a vision that started with the construction of the Owen Falls Dam (now the Nalubaale dam) and an eventual plan to 'tame the river Nile for the whole of its 3,850 mile-journey' (Hayes 1983, p. 331).[3] Not all felt as comfortable or confi-

[2] There are few formal historical accounts of infrastructure development in Uganda, let alone electricity, outside the work of Gail Wilson and the Economist Intelligence Unit (1957). Wilson undertook an examination of the electricity sector in Uganda in 1960 and 1961 for an MA thesis with the University of London. Her work was later published in a 1967 book titled *Owen Falls: Electricity in a Developing Country*. Wilson's account of the period leading to the construction of Owen Falls reveals some important details about the central concerns being debated at the time, most notably the escalating costs of the project.

[3] Hayes writes that the construction of Owen Falls Dam was part of a much larger concept that included the storage of waters in Lake Albert to the northwest and a massive canal to bypass the swamps of southern Sudan – a physical obstacle to the upstream southward navigation of the Nile. Hayes writes: 'The proposal therefore had obvious interest for peoples thousands of miles to the north of Uganda, and as a result of later discussions in Cairo it was agreed that the proposed dam at the Owen Falls would be constructed one metre higher than was necessary for hydro-electric purposes. This would make Lake Victoria Nyanza the world's largest self-renewing reservoir and would raise its level. All the water that Egypt would need could thus be stored and released, as required, at extremely low cost.

Figure 3 Letter of complaint to UEB

In any future correspondence
on this subject please quote

No. 25/5470

INTERNEE CAMP No. 6.

UGANDA.

23 APR 1945

April 21st 1945.

The Honorable the Chief Secretary
Entebbe.

CAMP LIGHTING.   *Copy filed in F.23/127/10*

    I have until now refrained from informing your of the many
failures of the electric lighting in this Camp which have occurred
in the past few months, because I realised that the East Africa
Power and Lighting Coy have had difficulties, but the position has
now become so serious that I feel I must place the matter before
you.

    Yesterday Friday 20th, I was warned by the Company that
light would be disconnected at 0800 hours in order to install the
new underground cable and that it was hoped to complete the work by
1700 hours. Light did come on in the Camp at approximately 1830 hrs
but by 1700 hours had gone off again. At 1930 hours after many
unsuccessful efforts to telephone the Company's local office I
finally rang up Mr. Paul in Kampala and was informed that the Cable
was faulty and had blown all lines right back to the Gaba Turbines,
thus putting the whole of Entebbe and Kampala into darkness. He
further stated that whilst he hoped to get light into both towns in
a short time it would be impossible to give light to the Camp since
the cable was the only available line and it had broken down. Thus
the entire Camp and the Veterinary Laboratory remained without
power or light through the entire night.

    During the month of April only the following breakdowns
have occurred :-

April 3rd. Complete breakdown at 0900 hrs. No light in
            Hospitals,Staff quarters,Veterinary Labs,etc.
April 4th. Breakdown continued until 1915 hrs when new
            transformer is installed.
April 6th. Breakdown 0300 hrs to 0905 hrs.
April 7th. Breakdown 0100 hrs - through night.
April 2oth. Light off from 0800 hrs all through night.
April 21st. 0900 hrs light still off.

    It is hardly necessary to stress the security reasons
which make the lighting of this Camp a matter of paramount importance but it might be as well to mention that the E.A.P.L. Coy
draws an income of over £. 1000. per annum for the lighting of this
Camp and that that fact alone should weigh with them in endeavoring
to see that the lighting shpply is reasonable efficient.

    The Veterinary Research Laboratory is also suffering from
these breakdowns and I understand that many important and urgent
experiments have been ruined completely owing to these lengthy
and frequent breakdowns.

Camp Commandant.

Copy to:
    Electrical Engineer.P.W.D. Kampala.

Figure 4 Letter of complaint to UEB

Transcription of Letter

3rd, March, 1945

The Honourable the Chief Secretary, Entebbe

I have the honour to attach a copy of a letter dated 15th Febryar, 1945 from the Deputy Coffee Controller concerning the effect of the faulty electric lighting system in Kampala on the output of Kampala Coffee Curing Works, and would be grateful for any action that can be taken to improve the present state of affairs.

S.T. Martin, Director of Agriculture

dent with the proposed plan as the two men. When 'moving the adoption of the Ordinance setting up the Uganda Electricity Board, the Financial Secretary referred to the scheme as an "act of faith"' (Wilson 1967, p. 2).

Uganda's first hydroelectric dam, Owen Falls

The decision to go ahead with the construction of the Owen Falls (now Nalubaale) Dam rested on an emerging belief that the provision of electricity was a service to be undertaken by the government and not merely a commercial enterprise. The factor determining the financial success of the new dam was quite simple: the electricity produced had to be consumed. However, the vision espoused in Uganda was also that new power schemes should be developed that could supply future potential demand in lieu of merely meeting existing demand (Uganda Electricity Survey 1947 in Wilson 1967, p. 2). In his 1967 book, *Development Projects Observed*, Albert Hirschman described this as the 'building-ahead-of-demand strategy' (Hirschman 1967, p. 68). Hirschman studied eleven development projects in 1964–65, one of which was Owen Falls. Regionally, this government-led approach to electricity development was a radical departure from the previous private, small-scale, distributed electricity systems. But globally this new vision was in keeping with emerging trends.

In their pioneering book, *Splintering Urbanism*, Stephen Graham and Simon Marvin investigate the relationship between networks of infrastructure and social and economic conditions in urban areas. The authors draw mostly from experience in the industrialized world but make important observations about the historic and contemporary role of infrastructure development in the Global South. In relation to the year the Uganda Electricity Board was created (1948), Graham and Marvin suggest that this was a period marked by the convergence of two broad phases in infrastructure development [the colonial (1820s–1930s) and neo-colonial (1940s–1980s) periods] and two 'styles' of infrastructure provision (2001, p. 81). In the first period, colonial governments had two objectives: (1) build infrastructure that would support the export of primary products; and (2) build infrastructure that would service local and colonial elites in order that they could organize production and exert political and administrative control (2001, p. 82). In the second period, these objectives were reinforced by the dominant development paradigms of the time – modernization and import substitution industrialization – which focused on the production and strengthening of industrial activities, by default in urban areas, to produce the assumed

(contd) The Egyptian government therefore agreed to meet the extra construction costs involved and also accepted the requirement for payment of compensation of lakeside dwellers whose land would be flooded or otherwise affected, round Victoria Nyanza' (Hayes 1983, p. 331).

trickle-down associated with the desired economic and social transformation. This was consistent with the evolution of electricity services in Uganda where EAP&L first serviced Kampala, Jinja and Entebbe with independent diesel generators in the mid-1930s to mid-1940s, under the assumption that urban areas and industry would be the chief recipients of electricity in future. So, while the decision to move ahead with a single large electricity generation source in Uganda – the Owen Falls Dam – seemed to be a 'leap of faith' to some, it was a leap that governments across the world were taking, including Britain.

According to scholars, critics and observers alike (Khagram 2004; McCully 2001; World Commission on Dams 2000), the 1930s marked the beginning of a period of global large dam construction, which escalated after the end of World War II. The early 1930s also marked the beginning of a period of advocacy for the construction of large dams, starting, in 1929, with the formation of a 'transnational professional association' made up of 'an array of engineers, builders, and bureaucrats' called the International Commission on Large Dams (ICOLD) (Khagram 2004, p. 6), which is still active today. But by this time, Britain was already well acquainted with the construction of dams both domestically and internationally, including large dams.

At the turn of the last century, Great Britain governed territories containing more than half of the world's big dams (Khagram 2004, p. 5). Britain had also completed the construction of the low Aswan dam on the Nile in 1902, which was subsequently heightened twice, reaching 36 metres in 1933. Further upstream on the Nile, Britain had completed the construction of the Sennar dam in Sudan by 1925 to provide irrigation for the Gezira Scheme, one of the world's largest cotton plantations at the time (McCully 2001, p. 18). Hence, the decision to move ahead with a large dam in Uganda was consistent with Britain's experience and vision, and well before construction had begun the colonial government was already planning heavy and secondary industry for the country (Hayes 1983, p. 332). In the words of Charles Westlake, 'Power from this scheme [Owen Falls] will make possible the liberation of the latent riches of Uganda. The industrial development will help to provide funds for education, training, housing and medical services' (Hayes 1983, p. 332).

Following the formal creation of the UEB two consulting firms – Sir Alexander Gibb & Partners and Kennedy & Donkin – were hired, and produced a report promoting Owen Falls as the first choice for Britain's first large dam in Uganda. Owen Falls was deemed to be superior to the two other locations being considered – Bujagali Falls (which was identified as a first choice in the 1920s) and Ripon Falls (which had been considered in the early 1900s) – given its accessibility, the potential to produce more electricity, the ability to better control the Lake's levels, and a sound geological base (Wilson 1967, p. 5). Despite Owen Falls' technical feasibility there remained much concern over the financial viability and cost of the project, particularly given the rising prices for capital goods and the devaluation of the sterling in 1949 (Wilson 1967, 3). Wilson explains that the technical construction of the dam was not going to be difficult, particularly given how well it was documented. Records of the level of Lake Victoria had 'been kept continuously since 1896 by the Physical Department of the Egyptian Ministry of Public Works. They confirmed that the flow of the Nile at Jinja was directly related to the level of the lake and was therefore predictable,' and there was 'no doubt that a steady and reliable flow would be available for the power station' (1967, pp. 4–5). Moreover, under the Nile Waters Agreement of 1929, Britain had agreed to obtain Egypt's consent for any development of the river. Hence, 'With minor modifications the Owen Falls Scheme became part of Egypt's scheme for "century storage" on the Nile. The dam was designed one metre higher than was necessary for the electricity scheme alone and it was agreed to restrict [part of the flow] in order to store water in Lake Victoria for Egypt. The Egyptian government financed the extra work and undertook to pay compensation for loss of power' (1967, p. 5).

The Owen Falls Dam was designed to have an ultimate capacity of 150 MW and was 'by far the greatest undertaking in Africa south of the Sahara' (Hayes 1983, p. 332). (It is noteworthy that, given the ongoing energy supply problem in Uganda, in 2006, at the height of Uganda's worst power crisis, the country was only generating 165 MW.) Generating equipment was ordered in July 1948 with arrangements for necessary labour made in the same year. At peak construction 2,500 workers were engaged – 2,000 of them African, 200 Europeans and 123 Asian (Wilson 1967, p. 6). Located on the west bank of the Nile (in the Bugandan Kingdom), Hayes writes that all were housed comfortably: 'Blocks of flats, bungalows, offices and stores, a new village, shops and recreational facilities were provided. Nothing like it had ever before been seen in East Africa' (Hayes 1983, p. 331). To finance the £400,000 construction of the estate the Uganda government floated loans in London. This effort certainly reinforced the initial evolution and boom of Jinja as a centre of industrial activity and as a point of migration for Africans and non-Africans.

In 1949, contracts were awarded to a consortium of private construction firms variously reported as being Danish, Dutch, British and Italian under the name *Owen Falls Construction Company*. Construction would continue for six more years, but by 1953, a year before completion, the financial records were showing a huge change in estimated costs. According to various accounts (Wilson 1967; Hayes 1983), project costs had reached almost three times the original estimate and twice the revised estimates that had been used in the decision to construct the dam. Hence, Hayes reports that by the time the dam was commissioned, 'the scheme had cost £14.7 million compared with Westlake's original estimate of £4.298 million ... Two loans, floated in London, had provided £12 million, and there was also the grant of almost £1 million from the Egyptian government' (Hayes 1983, p. 334). This is consistent with recent evidence suggesting that the mean cost overrun for large hydropower dams is upwards of 90% (Ansar et al. 2014). Wilson clearly explains that there were no dramatic problems causing the steep rise in costs; the scheme was executed at a time of rising prices and financial difficulties for the pound sterling (1967, p. 7). Nonetheless, because of these financial concerns the need to have an established customer base to consume the electricity *before* it was produced was reinforced. Hence, partly in response, in 1953 the UEB approached the EAP&L in Nairobi to ask whether it would be interested in purchasing a bulk supply of electricity when the dam was commissioned the next year. According to Hayes (1983), the answer was reluctantly 'yes' given the shortage of financial resources available due to conflict in Kenya at the time.

This decision also paved the way for an important historical event in Kenya's own history of electricity with the creation of the Kenya Power Company Ltd. The new company would become the purchasing agent for new power and an intermediate agent between suppliers and consumers in the Kenya. Hayes, in his history of the EAP&L, explains that the company had little choice but to agree to the bulk purchase of power from Uganda, during the period when the Governor of Kenya had just established the 'Emergency' prior to the Mau Mau period. Hayes notes that while the EAP&L had limited financial resources, it along with the Government of Kenya knew that if they wanted to develop agriculture west of Nairobi they would need 300 miles of transmission lines that they couldn't afford. Moreover, under its statutory limits, EAP&L could not raise the money required so the government, with the assistance of the Power Securities Corporation, established the Kenya Power Company Ltd to become the purchasing agent for new power and to 'interpose itself between the consumers and the suppliers of electrical energy and would be partly government-owned. By February 1953 the Kenya Power Company was formally registered in Nairobi as a private limited liability company' (Hayes 1983, p. 333). An agreement was eventually signed with Kenya in 1955 for the bulk supply of 45 MW of electricity for fifty years starting in 1958 at a fixed price of 2.9 cents per kilowatt-hour for the duration of the contract. From that point forward, Kenya was an essential customer of the UEB despite there being much concern in Uganda about the export of electricity given the shortage of domestic supply. In 1958, a third of Owen Falls output was exported to Kenya; by 1959 the figure had risen to 41.1%; in 1960, 44.2%; and in 1961, 47.8% of Uganda's output was sold to Kenya Power (Hayes 1983, p. 335).

The first turbine of the Owen Falls hydroelectric station began turning for tests in December 1953 with the official inauguration of the dam occurring the following year on 29 April 1954. It was a grand event. Queen Elizabeth II, only two years as the sitting monarch, was in Uganda to inaugurate the new dam (see Figure 5). The *East African*

Standard reported on the events of the day with full journalistic colour and intrigue noting how the 'concrete gleamed in the sun' (*East African Standard* 1954a). Another editorial began by expressing the same enthusiasm that the colonial government used to initiate the project:

> The principal ceremony today, the opening of the great hydroelectric undertaking at the Owen Falls on the Nile [will take place] near the spot … where [John] Speke stood and found the answer to one of the mysteries of the Dark Continent. Near that spot, half a century ago, the great leader of the Commonwealth and Colonial Empire, the stout-hearted defender of civilisation and human freedom, Sir Winston Churchill looked on the tumbling waters at the birthplace of the ancient river and visualized what has come about today. The Owen Falls scheme is a triumph of engineering achievement and of faith. It will be to East Africa a great power house, a symbol of inner light which Western Christian civilisation and the British people have lit in the minds of millions of Africans. It will give their lives a new direction and purpose for many generations. (*East African Standard* 1954b)

But as the editorial continues, the author goes on to raise important cautionary remarks, which, looking ahead, painted an ominous warning for the country.

> The scheme is based on the assumption that it will stimulate a great and prosperous change in the economy of Uganda, and in a measure of East Africa as a whole by making possible the development of the natural resources of the Protectorate and the industrial undertakings based upon the modern power which it will place in the hands of the civilized men. It has already been a more costly undertaking than was estimated and there is evidence that the conception of progress that brought it into being was more optimistic than has yet been realized, or seems likely to be achieved in the early future. The success of the enterprises which it is intended to serve depends on the availability of cheap power, on security of invested capital, on an adequate qualified force of workers, especially Africans, and on stable political policies and objectives which provide a guarantee of long-continuing conditions suitable and necessary for the evolution of an industrial revolution in a continent only yet emerging from its past into the light of the present day.
>
> If the Owen Falls scheme is to give its full value to Uganda and all its peoples, and if industry is to be attracted in adequate measure to justify it, these basic conditions of success must be a policy. The provision of the new source of power is only the beginning. Much hard-thinking – more than has already been given to the requirements of the difficulties – will have to be devoted to the implications of the policy of which the Owen Falls undertaking is the symbol if this great engineering feat is to contribute its full value to Uganda

Figure 5 Queen Elizabeth II visits Jinja for inauguration of the Owen Falls Dam (Photo courtesy of Uganda, History in Progress)

and its peoples and justify its existence and its cost. (*East African Standard* 1954b)

Charles Hayes points out that the Queen's remarks at the inauguration paralleled, with more reserve, the *Standard*'s enthusiasm and concern:

> In her speech inaugurating the project, the Queen said that the benefits of modern science had been brought to the enrichment of Uganda, to serve industries which were already in being and others which would be founded as a result of the availability of electric power. She went on: 'But let us not forget that economic development and the building up of industries are not ends in themselves. Their object is the raising of the people's standards of living. We welcome this great work because, by increasing the wealth of this country, it enables people – and above all, the African people – to advance ... I confidently believe that your children and grandchildren will look upon this scheme as one of the greatest landmarks in the forward march of their land.' (Hayes 1983, p. 333)

Today, even someone ill acquainted with the political history of Uganda will know that many of the conditions that were being articulated as necessary for the success of the scheme would not be realized, and the concerns expressed when initiating and inaugurating the dam were well founded. Starting shortly after Uganda's independence in 1962, the stability of the country and its policies would quickly unravel with a disastrous long-term impact on the state's infrastructure. Moreover, even in the years immediately following the dam's completion, when economic growth and electricity expansion increased, the expected economic and social transformation that was envisioned and espoused did not materialize. Indeed, during the inauguration ceremony, Westlake, Chair of the UEB, told a distinguished audience: 'Where electricity is abundantly available, progress in all fields of human activity inevitably follows. A country's state of development can be measured by the amount of electricity it consumes' (Hayes 1983, pp. 333–4). Given the legacy of low level of access to electricity in Uganda, clearly many factors inhibited the execution of this vision.

Politics and electricity in pre-independent Uganda: 1954 to 1962

By the end of 1954, three of the Owen Falls' generating sets were operating, with three others following in 1955 and 1956. The early intent of the scheme was to supply what Wilson describes as the 'modern sector of the economy, i.e. the towns and a few major industries' (Wilson 1967, p. 11). Initially, this meant supplying industries in Jinja and to a lesser extent Kampala, and 'the richer inhabitants of these towns' as well as 'Entebbe and later Masaka' (Wilson 1967, p. 9). The distribution systems in the towns were also to be compact and only cover the central areas (Wilson 1967, p. 9). There was no mention of electricity expansion and access to indigenous Ugandans initially. However, it quickly became clear that Uganda's industrial and urban consumer base could not support a project as big as Owen Falls (Wilson 1967, p. 11), despite the quick rise in the number of electricity consumers following the dam's construction. The UEB had failed to meet one of the conditions deemed necessary for the initial success of the dam – ensure that enough consumers existed before supply comes on line. As a result, 'the Board was forced to look for consumers wherever they existed, and to an increasing extent to carry its operations in to rural areas' where there remained long, unprofitable gaps marked by seasonality of consumption due to seasonal labour and agricultural activities (Wilson 1967, pp. 11–12).

Recognizing that the financial success of the Owen Falls Dam was dependent on adding additional consumers beyond those originally thought necessary by the UEB – industry and wealthy areas of urban centres – presented a new set of problems for the company: the pattern of rural settlement in Uganda was unfavourable to the supply of public utilities. People lived in isolated homesteads linked by 'winding paths and by-passed by roads' (Wilson 1967, p. 11). This pattern was and is fundamentally at odds with the least cost and most technically feasible layout of infrastructure – straight lines delivering utilities to densely populated regions where connections can be made quickly. Up until 1961, virtually all supply lines in Uganda followed roads, and therefore the pre-colonial and colonial settlement pattern provided little prospect for mass electrification. This observation is striking for another reason: it foreshadowed a tension that persisted in Uganda well into the 2000s and today. The settlement patterns in Uganda remain a formidable challenge to increasing the number of consumers in the country, and many people interviewed noted how much the low-density settlement patterns in Uganda inhibited rapid electrification (Lawrence Omulen, interview, 7 January 2003; Paul Maré, interview, 17 January 2003; Arthur Mugyenzi, interview, 20 March 2002; Thomas Tondo, interview, 13 April 2002). The settlement pattern was one reason that the 2001 *Energy for Rural Transformation* Project in Uganda was not focused on providing electricity to individual homesteads and villages but electrifying town centres where schools, health clinics and shops were serviced and therefore citizens too, indirectly.

The result, as Hirschman clearly noted in his study of the Owen Falls Dam, was that rural and low-income Ugandans were not going to gain access to electricity, in large part due to the high cost of expansion and low levels of consumption:

In Uganda, the national electric power agency ... undertook to build transmission lines to the various provincial towns or administrative centres as well as to coffee mills, cotton gins, and tea factories. But with power newly available in the towns and with transmission lines conspicuously transporting it overhead through the countryside, many nearby villagers thought that it would be a simple matter to supply them too and so petitioned the UEB. Some of them event went so far as to hopefully hang lightbulbs from their ceilings! Unfortunately, because the farmers' settlements were scattered and their prospective consumption very low, any large-scale extension of the distribution network into the countryside would be totally uneconomical and out of the question for the UEB which to this date [had] never turned a profit. Since, on the other hand, Uganda's so-called towns are little more than administrative and commercial centres almost exclusively inhabited by civil servants and East Indian traders, the UEB's transmission lines served essentially to make the rich and powerful more comfortable. (Hirschman 1967, pp. 62–3)

In a short period of time, then, owing to the pressure to make a profit, the electricity network expanded dramatically. In 1954 (the date Owen Falls was complete) there was 629 miles of electricity lines, but by 1961 there was 2,314 miles. Maps for these periods illustrate this change schematically (see Figures 6 and 7).

Figure 6 Electricity Distribution Network 1954 (Source: Wilson, 1967)

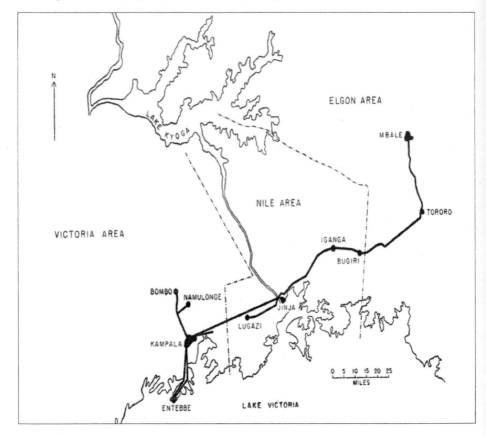

Prior to Owen Falls, the UEB extended the network for strategic and economic reasons in anticipation of the electricity to come from the dam. Hence, on the 1954 network map (Figure 6) the southeastern towns of Tororo and Mbale were both connected to the main grid. Tororo was important as it was the source of cement production for the dam starting in 1953, while Mbale and Iganga, like Tororo, were geographically easy to connect and had enough wealthy domestic and commercial (European and Asian) consumers to support supply. (The only areas supplied with electricity that had a dominant concentration of indigenous Ugandans prior to 1954 were located near Kampala's city centre – Katwe, southwest of the city centre, and Naguru in the northeast.) Following the dam's completion, the network was first expanded southwest to the town of Masaka, supplying the trading centres, missions and factories along the way (Wilson 1967, p. 15). In the town proper, the scheme supplied 'African commercial and residential areas, as well as the Euro-

Figure 7 Electricity Distribution Network 1961 (Source: Wilson, 1967)

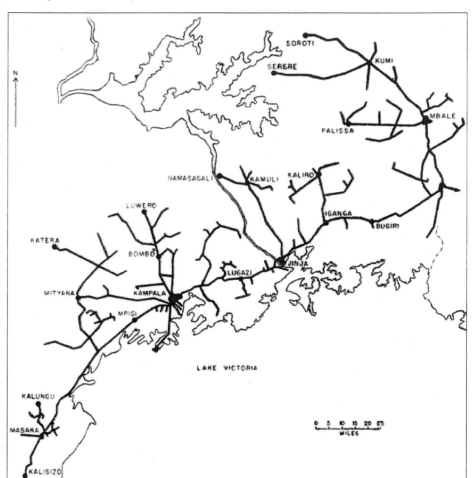

pean and Asian areas' (1967, p. 15). Masaka was chosen because it was one of the wealthiest coffee growing areas in the country, with four coffee factories nearby.

Up until 1960–61 transmission and distribution networks continued to feed off the main corridor between Masaka and Mbale, targeting industries and residential areas with consumers and areas with a high potential for new consumers. Even later, when the 1961 network map was compared to a 2015 network map, the main transmission network of the country has not changed dramatically with the exception of an extension from Soroti north to Gulu (see Cross-Border Information 2014). Hence, the quick expansion of the network up until the late 1950s was in keeping with the vision and most importantly the financial need of UEB. This quick expansion, however, also produced a difficult technical scenario for the UEB as they soon ran out of easily accessible customers.

Shortly after Owen Falls was commissioned in 1954, two members of the Economist Intelligence Unit came to Uganda on behalf of the UEB to inquire into the future development of the Ugandan economy. In the terms of reference for its report *Power in Uganda*, the Unit explained that they had four main objectives:

1) To prepare an appreciation of the probable course of economic development in Uganda, and of the stimulating effect which the availability of adequate supplies of electricity has had, and would have on the production of additional wealth in the Protectorate;
2) To estimate in general terms the growth in demand for electricity for all purposes in Uganda over the period 1957–1970;
3) To estimate in general terms the probable offtake [use] from outside the Protectorate;
4) To examine the probable economic and social effects if additional supplies of electricity were not made available from a second station. (Economist Intelligence Unit 1957, p. 1)

From these terms of reference, the Unit produced some very significant conclusions relating to the potential for indigenous Ugandan households to be connected, for expansion into rural areas, and ultimately, the need to start planning for a new dam immediately. The report notes:

- The outlook for electricity consumption in the industrial and domestic sectors ... rests on the assumption that electricity is in fact widely available;
- In spite of the very real benefits to African households which electricity confers, and of the expressed desire of Africans to become consumers, incomes are so low in most districts of the territory that *the mass of the rural population must be considered to remain unsupplied within the period under review* [1957–1970] ... the main domestic potential will probably lie in urban and sub-urban areas;
- The 1970 estimate for electricity consumption is 'significantly in excess of Owen Falls capacity, and would necessitate the operation of additional generation plant [sic] from a date well before this, probably as soon as 1965. *This means that construction work on a second dam will have to commence by 1960 and that preliminary survey work must be undertaken very shortly.*';
- There can be no doubt of the beneficial effects on an economy such as Uganda's of industrialization, urbanization and the improvement of domestic living standards. To all of these, electricity is an essential component and pre-requisite, and by economic stimulus it affords it

in turn creates new demand for itself. This factor, above all is likely to justify the construction of a second dam;

- 'Developments in India, North Scotland and Ireland, do lend support to the belief that *electricity extension schemes provide probably the shortest road to prosperity.*' (Economist Intelligence Unit 1957, pp. 7–9, emphasis added)

The report's conclusions provide a striking statement on the way electricity and its expansion were viewed. There was a clear articulation of the assumed relationship between electricity and prosperity, but a recognition that rural areas would *not* gain from the prosperity electricity might bring for a considerable amount of time. The UEB also felt that if prosperity was to be achieved, more generation capacity was imminently needed and that this generation would come from large-scale, centralized electrification schemes. The publication of the Economist Intelligence Unit report coincidentally, and perhaps fortuitously, coincided with the period of extensive growth and expansion of the electricity network (1957 and 1958 particularly). The report, therefore, helped feed the expansionist vision of the UEB at a point of high demand for electricity. But only a short time later, in 1960, these forward-looking plans would confront the reality that 'virtually all the potential industrial consumers in southern Uganda had either taken a supply of electricity or were close enough to the mains [central distribution network] to do so if they wished. The same was true of nearly all the gazetted trading centres in the more densely populated areas. The prospect for new consumers (though not for increased consumption by existing consumers) was therefore limited' (Wilson 1967, p. 19).

By 1960, then, the expansion rationale had quickly been met, and the Economist Unit's 1957 argument that 'how many industrial and domestic consumers can be connected up ... is a technical problem, rather than an economic or a financial problem' (Economist Intelligence Unit 1957, p. 3) was proving accurate. Despite this, coinciding with the findings of the report and in keeping with its swift action, the UEB commissioned Kennedy & Donkin with Sir Alexander Gibb & Partners to do another survey of the hydroelectric potential of the Nile in 1956–57. The report, *Report on Investigations on the Victoria Nile*, recommended not only a new dam at Bujagali Falls but also a three-stage development with two more dams downstream from Bujagali.

The preference for a dam at Bujagali Falls was based on earlier analyses of its potential and accessibility, but also a result of an emerging conservation ethic carried by the colonial government and emerging international non-government advocates like the World Wildlife Fund (WWF). Indeed, there was another hydroelectric site with three times the generating potential of Bujagali but, if chosen, would produce significant ecological and tourism consequences. The site with more generating capacity was located at Murchison Falls, in the heart of

Murchison Falls National Park, in the northwest of Uganda. The debate surrounding which site to choose for a second dam is notable not only for the environmental concerns it invoked, but also because it would be the first large electricity generation undertaking to be debated and initiated in the post-independence period in Uganda. For of course the expansion of the electricity network up until the early 1960s was taking place at the same time as rapid political change throughout Africa, and East Africa in particular. Indeed, it was a prominent member of the first political party that sought to reach out to Ugandans countrywide – the Ugandan People's Congress (UPC) – who would also preside over the decision to construct a second dam. The member was none other than future president Milton Obote.

But what stands out about the real and planned expansion of the electricity network prior to 1960 was that it evolved in general isolation from national, indigenous pre-independence politics. In fact, even the Economist Intelligence report and the subsequent hydroelectric studies done in anticipation of a second dam do not reflect on the national political context or environment and how this context would facilitate or negate UEB's vision. These were technical documents; UEB saw electrification as a technical undertaking. This context is important as it is consistent with the contemporary period of dam building in Uganda and one of the arguments of this book: contemporary dam building and electricity reform in Uganda evolved as if the political context was not changing or would not change.

Despite pre-independence political events that would leave a lasting legacy in post-independence Uganda – emergence of and competition between formal political parties, ethnic and religious conflict accentuated by colonial policies, the expulsion of the King of the Baganda (Kabaka), and constitutional negotiations – to read historic documents and reports relating to electricity in Uganda one would be forgiven for thinking that the British were going to carry on administering a docile Ugandan population in perpetuity as there is little reflection on national politics or the transfer of colonial administration.[4] On the eve of independence in Uganda (1962) and East Africa, however, the division between UEB's plans and national politics would start to vanish.

Electricity and independence: Conflict and network deterioration

By 1960, only five years after making the 50-year agreement to export electricity to the Kenyan Government, the UEB declared that it was

---

[4] The number of sources pertinent to a pre-independent political history of Uganda is numerous. For an authoritative and critical political history of Uganda that is frequently cited, see S. R. Karugire (1980). With respect to the legacy of colonialism on local government, see Fallers (1965), Burke (1964), and Mamdani (1996).

dissatisfied with the agreement. Uganda was concerned with the price and volume of electricity it had negotiated to supply for such a long period, while Kenya too had concerns with the volume of power it had agreed to purchase as the agreement was interfering with the development of its own hydroelectric resources; lenders, chiefly the World Bank, did not feel that additional domestic supply in Kenya was warranted. Hence, Kenya's energy independence was going to be compromised by the agreement to buy power from Uganda. Moreover, EAP&L, the agent executing the Kenyan side of the agreement, was frustrated with Uganda. Don Small, then head of EAP&L, felt Uganda was 'bogged down' and mired in long, drawn-out planning processes unable to execute on plans to increase electricity generation (Hayes 1983, p. 339). This frustration was made even more poignant given that EAP&L's Kenyan network had gone on expanding.

In the late 1950s and early 1960s, the Kenyan government had been investigating opportunities to develop its own hydroelectric scheme on the Tana River at Kitaru (Seven Forks) Falls. The plan was complicated because Kenya could not qualify for assistance from the World Bank as a 'colony' without British insurance. With independence on the horizon, the World Bank was, however, willing to consider supporting the scheme. It is noteworthy that at this time in Kenya indigenous African ministers were slowly being appointed, with one of the first being the new Minister of Power and Communications. Despite this, it was the EAP&L that had asked the Kenyan government to approach the World Bank. Using its own projections, the EAP&L plan was shown to the Bank at much the same time that Jomo Kenyatta became chief executive and Minister for Constitutional Affairs and Economic Planning (the office that held the EAP&L proposal).

The Bank's review of the proposal, however, was unfavourable, and it rejected the suggested need for the development of a new generating station in Kenya. The Bank believed that Kenya's domestic supply, combined with current and projected future supply from Uganda, was sufficient to support Kenya's needs. In the World Bank mission's report from 1962, it stated: 'We cannot see the justification for proceeding with the Seven Forks scheme at this stage and consider that it should be possible for both electricity undertakings to negotiate an increase in supply of power to Kenya from Owen Falls to their joint advantage' (Hayes 1983, p. 338). Hence, Kenya would enter independence on 12 December 1963 without the financial support to undertake a large independent electricity generation scheme while concurrently being forced to rely on supply from Uganda. The Bank's decision reinforced the Economist Intelligence Unit and UEB's desired network expansion and increased generation. But, like the UEB, it does not appear that the Bank considered how the political conflicts that emerged during independence in Uganda (9 October 1962) would affect future electricity supplies in Kenya.

The period between 1960 and 1962 was critical in Ugandan politics as political parties became much more entrenched, political demands affirmed and conflict heightened (Karugire 1980). With respect to electricity in Uganda, it was amidst this period of political change, and just prior to independence, that the Ugandan Government received a Specific Investment Loan from the World Bank under the title *Electric Power Development Project* (Power I). This project was part of the Uganda Electricity Board's $14.0 million expansion programme, of which $8.4 million was a loan from the World Bank. Hence, during Uganda's growing political conflict the UEB was staying its course and reaching out to the international institution it would rely on throughout its existence for financial assistance.

In 1964, shortly after independence, Prime Minister Milton Obote would oversee his first two of three initiatives relating to electricity in the country. First, the Uganda Electricity Act was passed re-establishing the Uganda Electricity Board as the sole provider of electricity in the country, with responsibility for generation, transmission and distribution to consumers. Under law, no other institution could play any role in electricity service provision. This situation would not change until 1999 when a new Electricity Act was passed. Second, a revised supplementary agreement with Kenya was signed to provide it with a bulk supply of 30 MW of electricity for 50 years. Despite these technical advancements, politically, the fragile power-sharing arrangement Obote had established to hold power was weakening. Mounting tension and widespread concern about corruption (Mugaju 2000) meant that by early 1966 the foundation for two decades of conflict and instability in Uganda were taking root.

In late February 1966, Obote made Colonel Idi Amin army commander. A week later Obote dismissed the President and Vice-President and assumed the functions of the Presidency. One month later, Obote abrogated the constitution and introduced a new 'revolutionary' constitution. In September 1967, Obote declared himself President, and abolished all political kingdoms in Uganda. (The Kingdoms of Uganda would not be legally restored until 1993 by an Act of Parliament and then institutionalized in the 1995 Constitution.) A short time later, in 1969, an assassination attempt on Obote would lead to all opposition parties being banned and leaders detained. But amidst this mounting domestic unrest Obote kept his hand on the country's electricity system.

Recall that just prior to independence the Uganda Electricity Board had two primary objectives – to find a new site to build a hydroelectric dam and to connect new customers. These two objectives were in some ways at odds given that the UEB was facing difficulties finding enough customers for its current electricity supply in Uganda, and some studies suggested that many citizens, particularly rural ones, would be unlikely to receive electricity service in the foreseeable future given the difficulty in extending the network. Despite this, considering the assumed relationship between electricity and economic development, UEB's vision, and studies suggesting that more generation was going to be needed in the near future, the development of a new hydroelectric site was a priority. Studies prior to independence identified six locations for future hydroelectric development. At the top of the list of sites with the most generating potential was Murchison Falls.

Located within Murchison Falls National Park, this tourist attraction was thought to be able to generate 600 megawatts of electricity – over three times that of Owen Falls. Furthermore, it was also located in the less developed north of the country, providing an opportunity to establish a generation source close to poorer northern populations, and limiting system losses due to long transmission distances from the south. While there were other promising sites in Uganda on the Nile – the other locations and generating potentials were: Bujagali (180 MW), Buyala (240 MW), Kalagala (240 MW), Kamdini/Karuma (246 MW), and Ayago (336 MW) – none had near the generating potential as Murchison Falls. However, none were also located in such a prestigious location either. If a dam were to be built at Murchison, it was estimated that 25 square kilometres of the National Park would be flooded and 90% of the river diverted, eliminating the sight Winston Churchill had described as 'the most remarkable in the whole course of the Nile' (Hayes 1983, p. 340). In addition, there was large concern that the dam would negatively impact a rich diversity of animal species in the park as well as conservation efforts relating to the rare white rhinoceros.

Conservationists argued that the development at Bujagali Falls was a preferable location to Murchison because a 'small' hydroelectric facility at Bujagali along with another one at Buyala – only two kilometres downstream from Bujagali – would produce close to the same volume of electricity as Murchison (Hayes 1983, p. 340). (Bujagali was not technically small according to the international convention for defining large dams – it was over 15 metres. But it was small in comparison to Murchison Falls.) In addition to the conservationist preference for Bujagali, history was on the side of the Bujagali site. The 1957 Kennedy and Donkin study of potential hydroelectric sites had also promoted Bujagali unless it could be shown that a 'large block of power of about the capacity of the full Murchison is required in the near future' (Hayes 1983, p. 340). Moreover, forecasts for electricity consumption in Uganda well into the 1980s were only half of what Owen Falls and Murchison dams would produce combined. Even in 2012, Uganda's peak generating capacity was 550 MW and peak demand was 489 MW (Ministry of Energy and Mineral Development 2012). Owen Falls and Murchison combined would have generated more than 800 MW in the 1960s. Had this come to pass it would certainly have tested competing theories about infrastructure development – to build ahead of demand or build and attract demand. Given that a scheme at Murchison Falls would have produced more electricity than most assumed possible to consume, in the UEB's 1965 annual report Bujagali was recommended over Murchison Falls as the next site for development. The Government of Uganda would approve this proposal the following year, leading the UEB to begin to look for funds overseas. One of the main sources of funding UEB turned to was the World Bank. But the Bank's confidence in the UEB was already low.

Given Kenya's earlier request for support to develop the Tana River for hydroelectricity, and now UEB's request for support to build a new hydroelectric site at Bujagali, the Bank suggested the consideration of a joint scheme. By 1968, joint consultations produced the 'Kenya–Uganda Coordinated Power Development Report'. The report provided figures and estimates of the costs of joint and independent development of hydroelectric schemes:

> If the countries went it alone Kenya would have to find $US 272 million and Uganda only $US 116. In a joint venture Uganda's contribution would rise to $US 175 million and Kenya's would drop to $US 185 million. Whilst the joint scheme (at $US 360 million) was cheaper to finance than would be the independent schemes (at $US 388 million), the co-ordinated scheme would require a 45 percent greater contribution from Uganda. (Hayes 1983, p. 341)

For Uganda and the UEB the choice of joint versus independent network development seemed obvious. Hence, a short time later, independent of Kenya, the UEB announced its decision to develop its resources on its own. To stoke its position UEB also announced that it was going to pursue the construction of the 600 MW Murchison Falls site *at the same time* as Bujagali. The UEB suggested that the surplus energy from Murchison Falls could be exported to eastern Zaire (DRC) and southern Sudan. As was noted in the beginning of this chapter and as will be noted later, the discussion and debate about how to proceed to increase electricity generation and for what purpose has dogged Uganda for decades.

In response to the suggestion that a dam in Murchison Falls National park would proceed at the same time as one at Bujagali Falls, the Uganda National Parks' executive officer, Francis Katete, renewed the argument against construction of a hydro facility in the park. With very similar arguments that would re-emerge thirty years later in relation to Bujagali, Katete stated: 'Both southern Sudan and eastern [Zaire] have formidable problems to overcome before they can be expected to provide a paying market for the sale of electricity ... Besides, the Board's [UEB's] current sales of surplus electricity to Kenya are worth half the price per unit compared to the Board's internal sales. To destroy a sure commodity (tourism) which nets good money (dollars, marks, pounds) in order to provide for dubious power exports does appear unjustifiable' (Hayes 1983, p. 341). According to Hayes, Katete's argument was that the UEB was 'throwing in the Murchison Falls site in the hope that it might appear rosier to international financiers'. But Katete suggested that conservationists would assist the Uganda Electricity Board if it decided to commission the Bujagali project. 'This would see the Murchison Falls Park survive into the next century, by which time, hopefully, other forms of electric power production will have been perfected to be competitive with hydro-electric generation' (1983, p. 341). It is once again striking how prescient Katete's remarks are.

First, in making the case for the protection of the National Park and its ecological resources, Katete critiques UEB's economic rationale for the project. Similarly, he asks why the Government of Uganda would want to forgo the guaranteed financial returns from tourism over the hypothetical returns from electricity exports. Second, Katete alludes to the potential of alternative energy sources being developed that will not require the development of Murchison Falls. As I will emphasize and elaborate in Chapter 4, what is most striking about these points is that they are nearly identical to critiques raised by non-government organizations in Uganda over the national government's decision to construct Bujagali thirty years later. It is also important to understand that Katete was not suggesting that no hydroelectric development take place. Indeed, he was suggesting that Bujagali still be built. This parallels Uganda in the 2000s as many non-government organizations clearly explained to me that their chief concern with the Bujagali project was not its ecological impacts but the absence of a process through which informed debate over the project took place.

Of course, in a short time the debate over hydroelectric development in Uganda would be eclipsed by civil unrest. Until being deposed by Idi Amin in 1971, Milton Obote advocated the Murchison Falls option. But Amin's reassertion of military rule, fear, political repression and civic unrest would mean that no Ugandan would see any further hydroelectric development for twenty years. Amin's infamous reign and demise (1971–79) would be followed by five administrations – the Uganda National Liberation Front, the Military Commission, Obote II, the Okellos, and, finally, the National Resistance Movement in 1986. According to one analyst, 'Between 1971 and 1986 there was no major development in the power sector' (Engorait 2005, p. 1), though Amin did temporarily cut electricity to Kenya in 1976 owing to a dispute over territory in western Kenya. The cut in supply represented 20% of Kenya's total supply (Hayes 1983). Some statistics will help illustrate the dismal state of electricity supply.

In 1968, the Owen Falls Dam was operating at full capacity, producing 150 MW of electricity (Engorait 2005, p. 1). By 1986, the generating capacity of the power station had degraded to 60 MW. In terms of consumers, Uganda's civil conflict also had a dramatic effect. In 1971, the year Amin took power, the total number of customers in Uganda was 69,500. And although the number of consumers increased during his rule, in 1979, the year the war with Tanzania ended, there were only

60,918 consumers in the country (Uganda Electricity Board 1996; 1999). The numbers of consumers recovered from 1979 to 1986, but the successive battles with Obote and the Okellos during the National Resistance Movement's armed struggle not only crippled the electricity system but presented a formidable national economic and social situation when it took control of the country.

Quantitative evidence of the collapse shows that between 1970 and 1980, monetary GDP dropped by 25% – equivalent to a reduction of per capita GDP by about 42%. By 1980, imports and exports had fallen by two-thirds from their peak value of 1972, industrial production had dropped by 80%, the number of vehicles and electricity consumption had fallen to two-fifths of their 1970 value, and state revenues had plummeted. Prices of local manufactures skyrocketed. Inflation resulting from low supply was aggravated by the emission of currency as a means of financing budgetary deficits. The cost of living for low-income workers rose by more than 500% between 1971 and 1977, while the minimum wage rose by 41% over the same period. Five years after Amin's fall from power, in 1984, real wages were less than 10% of their value in 1971. (Nabuguzi 1995, pp. 197–8).

By 1986, the number of electricity consumers in Uganda stood at 106,450. But two years later the number of consumers dropped again, to 80,795, largely owing to the poor state of the country's infrastructure. The essential problem that President Museveni and the National Resistance Movement were confronting was that the increased confidence in the stability of the country meant that businesses and individuals wanted access to electricity. However, the infrastructure was in such poor condition that demand far outweighed supply, requiring regular load-shedding, particularly during peak hours. And although a second World Bank-financed power project had been approved in 1985 to help rehabilitate the national system – Power II – the state of electricity infrastructure, access and provision was bleak.

Considering Africa as a whole, the continent generally struggled with the provision of energy resources from the 1970s onwards. In the 1960s, the development of energy sectors assumed that with an increased supply of petroleum and electricity, economic growth could be achieved (Davidson & Karakezi 1993). This misguided assumption (Tendler 1968, p. 17) and the development plans associated with it, were undermined in the early 1970s with the first rise in oil prices, and as we have read, through civic unrest. Excluding oil exporting countries, the cost of Africa's oil imports jumped from an average of 10% of export earnings to 20% almost overnight (Davidson & Karakezi 1993, p. 11). In conjunction with decreasing commodity prices, the increased cost of oil imports caused most countries to rely on external borrowing to pay for rising energy import bills. To respond to these events, countries restricted oil imports and established Ministries of Energy to try to address energy concerns and to coordinate government activities, but lack of clear objectives and appropriate structures made most efforts ineffective (Davidson & Karakezi 1993, p. 12). In conjunction with drought, concerns over the quality and availability of energy sources (oil, gas, fuel wood, charcoal) were becoming obvious; yet countries did not diversify their energy resource base.

The result was that countries were in no better position to handle the second rise in oil prices in 1979; oil import bills jumped from an average of 20% of export earnings to 50% for a number of-low income countries (1993, p. 12). With worsening terms of trade from continually falling commodity prices, the debt load of African countries increased dramatically. And even with the eventual decrease in oil prices, countries could not benefit given their weak economic conditions. Most energy utilities performed poorly in terms of revenues and increasing connections, not to mention that they lost or had little interest in renewable energy sources, which were not seen as programmes that could meet high demand. One notable effort to intervene in this crisis was the organization of the UN conference on New and Renewable Sources of Energy in 1981 in Nairobi, Kenya. However, with little financial support for the initiatives discussed, follow-up activities fell below expectations (1993, p. 13).

Conclusion:
History's influence on contemporary electricity and politics

The preceding discussion sheds light on the historic evolution and development of Uganda's electricity system and the broader debates about electricity access in the British East African colonial and post-colonial context. While some recent work has examined the interrelationship between colonial and post-colonial electricity supply and national politics (MacLean et al. 2017; Njoh 2016) there remain few studies of how pre-independence political and economic factors influenced electricity infrastructure. What can be gleaned from regional studies, primary documents and political histories of the region is that there is a clear and lasting influence of colonial policies and post-independence politics.

From a political perspective, what stands out in the case of Uganda is that during the period when the electricity network expanded most rapidly (roughly 1950 to 1960) and just prior to and following the completion of the Owen Falls Dam (1954), the Uganda Electricity Board seemed to function in relative isolation from the political events leading to independence. For example, I was unable to find any information on the relationship between the King of Buganda, the UEB and the provision of electricity or infrastructure to the kingdom and its government (an interesting observation given the long-standing relationship between it and the colonial government). What is more, UEB's and EAP&L's planning evolved in relative isolation from the tense political events surrounding them.

But what is perhaps most striking about this history is the degree to which the past challenges and debates about consumer connection and network expansion mirrored debates in Uganda forty to fifty years later. (As we will learn in the chapters ahead these include debates over public versus private led development, the appropriate scale of infrastructure investments, and the provision of service to rural and poor consumers.) Certainly, the degree of political and civil conflict that followed Ugandan independence plays a central role in explaining the degradation of the electricity system and low number of consumers. But the colonial government's assumptions about the intended outcomes of a rapidly expanded electricity network, along with the isolation of this development from the events surrounding pre-independence politics, did little to instill a smooth transition from colonial to post-colonial management of the electricity system. These observations are not helpful in providing solutions to Uganda's contemporary electricity challenges, but do, importantly, identify historical factors, which influence conditions today.

Four dominant themes emerge from this historical context, which resonate with contemporary debates about electricity in Africa and in Uganda's present situation: (1) electricity for industrialization versus individual welfare; (2) the role and influence of energy 'narratives' nationally and in the wider discussions of energy modernization or transformation; (3) the institutional legacy and weight of historic infrastructure investments or debates, particularly for dam construction; and (4) the role and influence of national political context on electricity expansion and availability nationally and regionally.

One of the interesting consistencies between historic and contemporary dam construction efforts in Uganda is the dual challenge of trying to predict the amount of electricity needed and the number of potential consumers available to pay for what is produced. This was certainly the issue the UEB encountered in 1960 when it found that its quick connection of businesses and European and Asian consumers (along with some African consumers in the urban centres and large trading centres in the southern portion of the country) left it needing to look to rural areas and 'Africans' for additional consumers. In Chapter 4, when the debates surrounding the construction of the Bujagali dam are examined, this issue will again reveal itself as questions about the potential to consume all the electricity produced by Bujagali, along with the cost of the electricity produced by Bujagali led some observers to call into question the Ugandan government's linking the construction of the dam to more individual consumer access and poverty alleviation. In the pre-independence period in Uganda the issue was never that citizens did not want to be connected to electricity. The issue was whether the managing authority could expand the network technically and financially, and whether consumers, more specifically Africans, could afford to connect to it. The 1957 Economist Intelligence Unit report noted that '[i]n spite of the very real benefits to African households which electricity confers, and of the expressed desire of Africans to be consumers, incomes are so low in most districts of the territory that the mass of rural population must be considered to remain unsupplied with the period under review [1956–70] ... [and] domestic potential will probably lie in the urban and sub-urban areas' (1957, p. 8). The report continues by stating that one of the 'fundamental features of the Ugandan economy [which] may be regarded as providing the "determinants" of the future growth in sales of electricity to African households' is the 'Africans' willingness to consume, and pay for, electricity' (1957, pp. 121–2). Given the higher relative value 'the African' places on electricity as compared to 'the European', the report notes, 'the price which the African can be persuaded to pay for these goods may also be much greater than expected, in view of his relatively low income' (1957, pp. 121–2). Hence, the report concludes, 'it is evident that the key to long-term expansion in the domestic sector rests with supplying the maximum number of African households' (1957, p. 133), despite the challenge that rural settlement patterns place upon this potential drive. These historic remarks highlight one of the central debates in sub-Saharan Africa and contemporary Uganda: is energy justice possible? Is an equitable distribution of electricity quality and supply possible? Is electricity a service that the poor should be provided with, or is it a luxury good, which should be acquired only when it can be afforded?

In the 1957 Economist Intelligence Unit report the answers to these questions were clear. The rationale for expanding electricity was not based on right or individual need, but on corporate financial need and opportunity, and visions for national economic growth. Individual and national prosperity from electricity access was recognized, but the real drive to expand electricity access was to generate revenue and consume the electricity generated. As the Economist Unit notes, 'the African' desired electricity, but connecting people without their being able to pay or subsidizing the service was not considered.

But what is particularly striking about these historic arguments about 'willingness to pay' and electricity for industrialization is that the path to development through electrification was assumed to be indirect. Electrification would foster industrialization, waged labour and increased domestic savings, which would then enable households to afford electricity. Household electricity provision was not viewed as development; enhancing the capabilities of households was not deemed to be development. This line of thinking is not surprising given that the high period of dam construction and electrical infrastructure expansion in Uganda coincided with the golden era of economic modernization theory.

However, when thinking about the implications of this in 2017, it is important to consider how this historic vision of electricity as a tool for economic development carried forward in Uganda. Has the rationale for large-scale infrastructure development been based on the same principles espoused historically? These questions point to the second theme – the presence of a dominant energy narrative in Uganda.

As earlier stated, a development narrative is often framed as having a beginning, middle and end, or premise and conclusion, and 'revolves around a sequence of events or positions in which something happens or from which something follows ... development narratives tell scenarios not so much about what should happen as about what will happen – according to their tellers – if the events or positions are carried out as described' (Roe 1991, p. 288). Accordingly, it is reasonable to assert that an energy narrative in Uganda started to emerge and be communicated in the 1960s. This narrative was first firmly articulated by Winston Churchill in the early 1900s and then carried on by the Uganda Electricity Board. What should happen was quite simple: make electricity available for industry and economic development and more demand will follow. As the dominant 'narrator' of the story, UEB was able to present a convincing vision about what would happen in Uganda if the Owen Falls Dam and subsequent dams were built. Supporting this narrative was research from international consultants and the World Bank, as well as financial support from the World Bank and the UK's Colonial Office. The mechanism by which this narrative was to be achieved was the Owen Falls Dam.

But as was revealed, this approach quickly ran into problems. Soon after the Owen Falls Dam was complete and the network expanding, the technical, social, economic and political reality of the country challenged the story of what was supposed to happen, revealing that the knowledge or vision feeding the narrative in Uganda was complicated by many other factors. As Hirschman explained, this meant that while UEB needed more consumers it could not afford to expand the network to reach them, leaving electricity for those already wealthy or in positions of authority (Hirschman 1967, pp. 62–3).

> The UEB stood ready to bring power to the villages in the vicinity of the towns it supplied, provided the villagers made an adequate capital contribution to the cost of the transmission, step-down transformers, and distribution. But since power was brought to the towns (and therefore to the East Indians) wholly at UEB's expense, this policy was resented as rank discrimination against Africans. (Hirschman 1967, p. 63)

The cornerstone of this emergent energy narrative was the link between electricity and economic development and modernization – a belief that electricity for industrial activity was to take place *before* individual access and that individual provision, particularly to Africans, was to be done only out of financial necessity for the company or if the consumer could afford it. Conflict over this kind of energy narrative remains prominent. Historical evidence shows that indigenous Ugandans were prioritized as consumers of electricity only in the context of expanding the network, and only if they could afford to pay for the full or indeed a higher price for the service. Hence, there is some important resonance between UEB's ambitions and approach in the late 1950s and early 1960s and the Government of Uganda's vision for electricity development in the late 1990s and early 2000s.

Equally, the relationship between the dominant narrator of the energy story in Uganda in the colonial and early post-colonial period and other interests remains prominent. These interests historically and today include international consultants, private firms, the World Bank, local governments, the national government, international finance capital, and domestic and international non-government organizations. The focus on narrative also brings in the debate over who will lead development – the public or private sector. This aspect of the energy narrative is significant as the pendulum in Uganda began with the private sector, then swung to the public sector, then back to the private sector in the late 1990s, and now has swung again back to the public sector.

A third dominant theme from this history relates to the legacy of the Bujagali site. Historical documents and research reveals that Bujagali was identified as a prime site for the construction of a hydroelectric dam in the early 1900s. In fact, in the 1920s it was identified as the preferred location for a dam in Uganda but, owing to easier access to the Owen Falls site, was downgraded to a second or third choice. Hence, in the context of the debates surrounding the appropriateness of Bujagali as a site for a hydroelectric dam, opponents to the project needed to recognize the historical weight or legacy that the Bujagali site carried in the overall plan for electricity development in the country. Bujagali existed on paper and in the institutional history of electricity in Uganda for almost one hundred years. As one interviewee explained to me, even if Bujagali was not deemed immediately appropriate for development it will always exist in the minds of government and consultants given its formal presence in historic documents and reports.

For political science, arguments about the 'weight' of historic decisions on future decisions are framed under the notions of 'historical institutionalism' and 'path dependency'. At the heart of this analysis is the theory that 'each step along a particular path produces consequences which make that path more attractive for the next round. As such effects begin to accumulate, they generate a powerful virtuous (or vicious) cycle of self-reinforcing activity' (Pierson 2000, p. 253). Pierson writes:

> This conception of path dependence, in which preceding steps in a particular direction induce further movement in the same direction, is well captured by the idea of increasing returns. In an increasing

returns process, the probability of further steps along the same path increases with each move down that path. This is because the *relative* benefits of the current activity compared with other possible options increase over time. To put it a different way, the costs of exit – of switching to some previously plausible alternative – rise. (Pierson 2000, p. 252)

For development scholars, Hirschman's notion of the 'Hiding Hand' parallels contemporary discussion of path dependency but by focusing directly on the factors influencing individual institutions or decision-makers.

Hirschman used the notion of a Hiding Hand to symbolize the invisible or hidden hand that conceals project difficulties from decision-makers until the process is well under way. The principle suggests that project planners often underestimate the costs of projects knowingly and unknowingly, and when confronted by the difficulties in implementation that arise during the process, must push harder for the project to be completed. Inevitably, when advocates push 'harder', conflict increases. Hence, as a third theme, it is important to recognize how history led to a large hydroelectric project in Uganda – Bujagali – becoming a first choice for solving electricity problems in the country, but while the regional and national challenges and lessons from past undertakings did not receive due consideration. These factors point to the last theme that needs to be to carried forward into the remainder of the book: the role of politics in decision-making and the contest between different interests in decision-making.

As the above historical discussion revealed, there are few studies that implicitly or explicitly analyse the politics of infrastructure and electricity in East Africa.

What can be understood is that up until independence, decisions over infrastructure provision were largely independent of the social, economic and political reality of countries, particularly in Uganda. For a short time, Milton Obote became involved in the debate over a second dam, but this was short-lived after being deposed by Idi Amin. Under Amin, the electricity network deteriorated, and perhaps his only notable interest in the network arose when he used electricity to sanction Kenya, when it is implied that he instructed the Uganda Electricity Board to interrupt supply over a dispute about territory. Hence, given the legacy of conflict in Uganda for two decades (the mid-1960s to the mid-1980s) and the fact that no large-scale electricity generation source was constructed between 1954 and 1993, history reveals that successive national governments showed little interest and/or had little ability to influence the expansion and construction of electrical infrastructure. This is despite the Uganda Electricity Board making efforts to maintain and expand the system during these periods while non-government interests did have influence over decisions.

The original expansion of the network in Uganda was dependent on an $8.4 million loan provided by the World Bank's International Development Association (IDA). This loan, the Electric Power Development Project (Power I), was approved in 1961 and was the World Bank's first project in Uganda. Regionally, and in contrast, Kenya's inability to develop one of its own hydroelectric sources during this period was due to the Bank's argument that there was enough capacity in Uganda to suit Kenya's needs. Hence, the World Bank's role in infrastructure and electricity in the region is longstanding, and deeply influential. The Bank's decision not to finance Kenya's development of the Tana River hydroelectric scheme made Kenya dependent on Uganda's electrical resources while also increasing the imperative of further developing Uganda's hydroelectric resources. Indeed, while the World Bank's general influence and importance in Uganda in the 1990s and early 2000s is well known, the influence it had over government decision-making – whether implicitly or explicitly stated – is central to electricity in the country and region.

Other non-government interests also had an important historical influence, from Britain's early decision to designate Uganda as a Protectorate to secure access to the Nile waters; to the consulting firms used to study hydroelectric development options on the Nile; to engineering firms used to construct Owen Falls Dam; to the international capital raised through bonds to build the dam; to the small but interesting early influence of international environmental NGOs in debates of dam site selection. The influence of external interests on Uganda's electricity infrastructure proved significant historically and, as the chapters that follow show, remained significant in the debates and actions intended to facilitate an energy transformation in later years. Hence, Uganda's history with electricity expansion and development reveals multiple, multilevel conflicts and transformations in the pursuit of new energy pathways.

3

Privatization &
Electricity Sector
Reform

From the late 1960s until the late 1990s, power industries in Africa were most often national monopolies in charge of providing a public electricity service (Girod & Percebois 1998, p. 22). The three segments of electricity service delivery – production/generation, transport/transmission, and distribution – were vertically integrated, with supervision and regulation supported by public ministries or quasi-independent regulatory agencies. The rationale for this arrangement stemmed from the belief that public utilities had to support national development and the cohesion of society through the distribution of an important public good – electricity (1998, pp. 22–3). A fall in sales of electricity during the 1980s following economic decay in sub-Saharan Africa, however, left most utilities unable to expand or provide consistent service and unable to maintain equipment and infrastructure. The fallout from this era carried forward into the 2000s with large system losses in many countries due to technical problems and poor quality infrastructure. A great deal of blame for electricity sector problems can be attributed to dismal economic conditions and financial constraints, but problems were also frequently attributed to the quality of administrative oversight of the sectors.

Beginning in the 1980s, private sector proponents argued that public utilities in Africa lacked internal motivation, had little management autonomy, and were vulnerable to political interference (Girod & Percebois 1998; Yi-chong 2006). It was also argued that public utilities were not motivated to look for greater efficiency; were able to transfer costs resulting from poor management to the national pocketbook; and were able to finance investments using other government funds (Girod & Percebois 1998, p. 24). This situation prompted calls for restructuring, reorganization and the general reform of electricity sectors to improve and increase service delivery (Davidson & Sokona 2001; ESMAP 2000; Girod & Percebois 1998; Turkson & Wohlgemuth 2001; UNCHS 2001; Wereko-Brobby 1993); it also led to the promotion and creation of new organizations and institutional incentives that would promote substantial investment in, and expansion of, electricity infrastructure (UNCHS 2001, p. 143), while at the same time reducing political interference in the management of the sector.

The World Bank's promotion of electricity sector reform was central in the 1990s and early 2000s: it was 'the main architect of energy sector reform and liberalization' in developing countries (Vedavalli 2007, p. 78). In general terms, the World Bank advocated for change under the suggestion that an 'accountability framework for service delivery' was required (World Bank 2004). In promoting this approach, the Bank highlighted several network utility and electricity- specific reform actions, such as: obliging enterprises to operate according to commercial principles; restructuring the power supply chain and introducing private competition in order to improve efficiency, customer responsiveness, innovation and viability; introducing transparent regulation that is independent of government and electricity suppliers; and focusing government's role on policy formation and execution and divesting from generation and distribution (Bacon & Besant-Jones 2002, pp. 3–4).

At the centre of these suggested actions was a vision that increasing the role of private, for-profit companies in electricity provision would be beneficial. But when this argument is compared to the historic way that countries developed national infrastructure systems, the uniqueness of this proposal is revealed.

When industrialized countries developed their infrastructure networks they most often relied on a vertically integrated model of electricity service delivery (see Graham & Marvin 2001). Recognizing this contradiction, the World Bank asked: why should developing and transition economies take on this new approach? (World Bank 2004, p. 4). 'The simple answer', the Bank wrote, 'is that the new model, *implemented correctly*, offers benefits too big to ignore – for governments, operators, and consumers. The primary virtue of unbundling is that it promotes competition, ensuring that firms provide their services at efficient prices' (ibid, emphasis added). The emphasis on 'correct implementation' signalled that some ideal existed, but when conflict, debate or impediments to the 'correct path' emerged, what was the outcome? What, then, is the record of electricity sector reform in Africa? What are the conditions that were promoted for reform success, and did these conditions match the reality in African countries? Identifying the arguments supporting reform, along with the domestic in-country conditions and processes that were expected to make reform successful, provides an opportunity to compare 'theory with practice'.

This chapter examines trends in electricity sector reform that emerged in the early 1990s in sub-Saharan Africa. It focuses particularly on the motivations for reform and how those reforms corresponded to the general conditions of the electricity sector and access. The chapter ends by highlighting how these reform trends materialized in Uganda and what precedents they established as the country moved forward with sector reform in the 1990s. The chapter reveals how reform discussions in the 1990s generally focused on fixing what were perceived to be technical problems with the sector – administratively, financially, and from an infrastructure perspective – but did not consider how or if these recommended reforms corresponded to the evolving social and political context in the country.

Electricity and privatization in Africa

In sub-Saharan Africa (SSA), many countries remain in the midst of electricity sector reforms. Even the countries with the largest economies – South Africa and Nigeria – continue to struggle to implement reforms that will meet demand. For the sub-continent, research on the impacts of reforms has remained weak and has depended heavily on data and lessons from transitional economies of the former Soviet Union and Eastern Europe (Birdsall & Nellis 2003, p. 1627). Given how recent reforms have taken place and the continuing sector problems in many countries, there is still very little known about the impact of various reform efforts and models in Africa. Further, there is very little knowledge about the impacts of reform on social welfare, with most attention paid to changes in levels of access to electricity over time. Research that has been done tends to provide broad overviews of several-country experiences or in-depth, rich understandings of single-country challenges with electricity provision (Olukoju 2004). More recent collections have presented valuable lessons about different reform models and regulatory structures, which is a welcome contribution (Kapika & Eberhard 2013). These findings, however, reveal how difficult and long-term electricity sector reform processes in African countries are: reforms have only progressed partially, with the 'envisaged end-state of the standard reform model unlikely to be reached for the foreseeable future … fledgling independent regulatory authorities have been forced to grapple with regulating power sectors structured in a manner that was not envisaged when the standard model was first advocated' (Kapika & Eberhard 2013, p. 6).

In the 1990s, it was generally known that countries were experimenting with different models: at one end of the reform spectrum there were countries that had privatized public companies completely (Côte d'Ivoire, Guinée and Mali); at the other end there were countries that had maintained a predominantly vertically integrated monopoly (Angola, Botswana, Eritrea, Ethiopia, Malawi and Niger). In most other countries, some type of reform had occurred to facilitate vertical de-integration and/or contractual service arrangements to encourage and permit private investment, and to promote various degrees of private competition. A long list of countries fitting this category included: Benin, Burkina Faso, Burundi, Cameroon, Congo, DR Congo, Ghana, Kenya, Madagascar, Mauritius, Mozambique, Namibia, Rwanda, Senegal, South Africa, Tanzania, Togo, Uganda, Zambia and Zimbabwe (see Girod & Percebois 1998; AFREPREN/FWD 2005). Hence, most countries south of the Sahara were engaged in some type of sector reform but with little evidence of reform impacts (Wamukonya 2003, p. 1282).

Historically, utility reform efforts in Africa concentrated on internal organizational change owing to financial problems, and associated administrative concerns (Nellis 2003). In turn, it was common that the IMF would encourage cuts to budgetary supports for state enterprises as debts were incurred but not serviced (ibid). Following the IMF's identification of the problem and insistence for improvement, the World Bank became more involved in terms of the design of reform and privatization, and implementation. 'In many, probably most African countries the principal motivation for privatization has been to placate IFIs [international financial institutions]' (Nellis 2003, p. 6). This is not an ideological position; it is supported by former World Bank employees intimately involved with energy (see Vedavalli 2007). By the early 1990s, 'industrialized countries, multilateral institutions such as the WB and the IMF and NGOs [the World Energy Council] ... began to emphasize the inevitability of developing countries to adopt a free market system and to liberalize their economies to facilitate public and private investment in energy' (Vedavalli 2007, p. 56). The Bank and its borrowers believed that they could not keep using a 'business-as-usual' approach to lending when power utility performance was deteriorating; its role was to facilitate and require 'developing countries to pursue pricing and institutional reforms to attract private investment' (Vedavalli 2007, p. 56).

The turn to the private sector was not simply an ideological conviction; it followed the wave of liberalization of electricity markets in the US and UK, as well as other countries like Germany, Poland, the Czech Republic, France, Austria, Sweden and Hungary (Vedavalli 2007, p. 30). Moreover, the promotion of the private sector was intended to facilitate desperately needed investment in infrastructure in developing countries generally, and Africa specifically. What seems to have been missing from the promotion of private sector participation was an assessment of the willingness of the private sector to operate in African countries, and the scale of institutional reform and political change that would have to precede or parallel this participation. Financial and political risks in countries like Uganda, with low levels of electricity access and poor infrastructure, meant the list of 'ideal conditions' for private participation was difficult to attain.

To overcome the fears inherent in widespread reform while still providing room for investment, most African governments – whether by choice or by requirement – accepted that sector reform would entail private firms playing a central role in service provision – the central issue was at what pace this change would be introduced and to what extent the state would continue to play a role in service provision. Thus, independent power producers (IPPs) were invited into electricity and other utility sectors with the expectation that they would help expand services, construct new facilities and improve operational efficiency. But in contrast to other regions, poor network infrastructure quality, low electricity supply, low levels of connections and high poverty meant that there was low or extremely cautious investor interest in African network utilities in the early days of private sector promotion. As a result, in addition to changes in administration, regulation and tariff structures, strong financial incentives were needed to attract private firms (Bayliss 2002, p. 6). Sometimes lack of investor confidence became a major stumbling block in reform and privatization: 'Transactions have been painfully slow. Enterprises which have been in a limbo state of "being privatized" for several years have rapidly declined' (Bayliss 2002, p. 6).

Historically, the strongest resistance to privatization has come from within state bureaucracies owing to reductions in numbers of employees and sometimes also salaries (see van der Walle 1989; Batley 2004). The speed at which privatization was promoted, and the manner by which it was implemented, also produced discontent, with the social impacts from utility privatization being much less favourable and less known, and with improvements in efficiencies rather than equity (Birdsall & Nellis 2003, p. 1623). With respect to efficiency, administration and finance, research suggests that positive outcomes can occur over time with utility privatization: private owners receive good financial returns, there is improved technical and operational efficiency, and the state reduces its administrative and financial obligations (Birdsall & Nellis 2003). In addition, there is some evidence that over time, network expansion and increased access to services to the urban poor does occur, with less favourable results for the rural poor, who are often left out or households far from the main grid. Analysts and advocates of privatization also suggest that these outcomes, combined with a demonstrated commitment to reform, can help sustain a larger process of market-enhancing economic reform in countries (Birdsall & Nellis 2003; Centre for Global Development 2003; Komives et al. 2001; World Bank 2003). In some countries, civil society organizations' resistance to privatization has also been prominent, particularly in countries with large and prominent labour unions like South Africa and Nigeria.

The World Bank, among others, acknowledged the global anti-privatization sentiment that existed at a time of price increases, job reductions and the high profits of firms that had improved operating performance (Birdsall & Nellis 2003; World Bank 2004b). 'But these adjustments', the Bank wrote, 'have been necessary for privatization to achieve its public interest objectives' (World Bank 2004b, p. 6). Nonetheless, to address the range of distributional problems that arose with privatization, some researchers noted that greater attention to the process of reform needed to be promoted: governments, and those that assist them, Birdsall and Nellis wrote, 'should invest more upfront attention and effort in the creation and strengthening of regulatory capacity, and less in organizing quickly transactions. This means taking the time to lay the required institutional foundations' (Birdsall & Nellis 2003, p. 1628). Further, governments should not ignore equity problems, assuming they are unavoidable and the 'temporary price to be paid when putting assets back to productive use' (ibid, p. 1629). Indeed, in addition to the economic and social benefits of a well-designed reform process that is cognizant of distributional impacts, there are also potential political benefits. Minimizing and countering the real and perceived unfairness of privatization 'is worthwhile, so as to preserve the political possibility of deepening and extending [future] reforms ... a democratic government cannot implement reform when masses of people are in the streets attacking that reform, and, of course, no government can enact reform if it is not in power' (Birdsall & Nellis 2003, p. 1629).

Hence, while a new model of service delivery was being promoted in Africa, people were warning about the need to be conscious of the political implications of the pathways of reform chosen. These warnings echoed Hirschman's observations in the 1960s about the possible negative outcomes of ignoring the indirect effects of a poor project process and more recent observations about problems that arise when decisions are made based on technical merits rather than the social and political character of countries or project settings (see Easterly 2013).

Taking the political and social context into consideration is not a recipe for reform or privatization success. Further, it is not possible to do a post-hoc analysis of failed reforms to say that if reform had been slower or been more cognizant of political and social contexts reforms would have been better. For example, as I discuss below, the problems in Uganda's electricity reform process in the early 2000s most certainly affected the quality of service and accessibility of service a decade later. But it is not possible to show a direct correlation between the problematic process and electricity outcomes owing to the many overlapping problems that occurred. Hence, the point of highlighting the significance of politics and process here is to bring attention to the fact that these concerns are not new and that, whether directly correlated to the outcomes or not, promoting privatization and complex reforms without considering the real or potential social and political change in transitional democracies and the effect of these changes on the population can inflict direct and indirect penalties on governments and citizens. Given that international agencies and research formally cautioned against reforms purely based on technical goals, the question remains how social and political conditions were considered in the practice of reform.

Technical and financial matters are of course critical to infrastructure outputs. But how these technical requirements intersect with the political and social context remains a critical concern no matter what model is selected. Indeed, as will be highlighted later, it was through formal political processes (not street-level protests) that domestic civil society groups in Uganda challenged the electricity sector reform process and associated dam construction efforts: civil society organizations began using the formal institutional processes and structures that were already in place, such as the courts, to challenge government programmes rather than taking to the streets. These actions, along with legislative review processes, exemplify the way that utility reform processes offer a window into a changing political environment in sub-Saharan Africa in the late 1990s, whereby legislative systems played an important role in their democratic evolution (see Barkan 2009), including in Uganda (Kasfir & Twebaze 2009).

If the contemporary and historical experience with utility sector reform and privatization reveals a tension between technical service delivery goals and political and social considerations, to what extent were frameworks or guidelines promoted by international organizations reinforcing these tensions? How were technical, institutional and regulatory principles engaged with procedural and political ones?

Theory v. Practice: Ignoring conditions for reform success

As one of the central advocates of competition and private sector participation in Africa, the World Bank's publications and advice serve as an important reference point when examining reform. In a 2002 World Bank Energy and Mining Sector Board Discussion Paper, four principles for successful electricity sector reform were identified: (1) the formation and approval of a power policy that provides broad guidelines for the sector; (2) the development of a transparent regulatory framework; (3) the unbundling of the integrated structure of the power supply; and (4) divestiture of the state's ownership, at least for generation and distribution (World Bank 2002, p. 4). Later, the Bank further debated the merits of reform and private sector participation, while reasserting principles that make reform, unbundling and private sector participation successful (World Bank 2004b, pp. 4–8). While regulatory efficacy was highlighted again, proper sequencing of reform was also emphasized (World Bank 2004b, p. 8). Table 1 summarizes these principles.

These principles established a framework for reform, but they were not a recipe for success. The Bank, in fact, offered cautionary remarks about restructuring in some reports in the early 2000s, which emphasized the need to consider domestic conditions:

> There is no universally appropriate model for restructuring network utilities. And the fact that state ownership is flawed does not mean that privatization is appropriate for all infrastructure activities and

Table 1
Guiding principles for energy sector reform, unbundling and privatisation

Guiding Principle	*Description*
Formation and approval of a power policy	As a first step, a power policy is needed to provide direction to the sector and to sector reform. It is assumed that it will articulate the direction of reform, including priorities and sequencing.
Proper sequencing	Sector restructuring should be in place prior to the entry of private firms so that regulation and organizational responsibilities are defined, and sector oversight and management is functioning.
Transparent regulatory framework and efficacy	Clear regulation and regulatory authority is needed to define roles and responsibilities of various sector actors. In the absence of clear regulation investor interest may be reduced and/ or the potential risk of entry high. Rules and the authority overseeing rules governing the sector should be finalized and established prior to unbundling and private sector entry.
Unbundling the vertical structure of power supply, divestiture of state's ownership, and institutional restructuring	The monopoly power company should be unbundled to create separate independent companies (generation, transmission, and distribution). This should be followed by the last step in privatization, which is to divest from the newly independent state-owned companies.
Proper pricing	Prices should reflect 'the real cost' of service provision, including investments in maintenance and expansion of service delivery. Subsidies should be carefully employed in order to maintain the necessary 'revenue base'.
Secure private investment	Private sector capacity to take over unbundled commercial enterprises (if being pursued) should be secure and guaranteed. Contractual agreements and rules must also be clearly established, with mechanisms in place to ensure application of rules and amendment to rules.

(Sources: World Bank 2004; Bacon and Besant-Jones 2002; Vedavalli 2007)

all countries. Before state ownership is supplanted by another institutional setup, it is essential to assess the properties and requirements of the proposed alternative – taking into account the sector's features (its underlying economic attributes and the technological conditions of its production) and the country's economic, institutional, social, and political characteristics ... (World Bank 2004b, pp. 8–9)

One key domestic factor that was considered important for electricity reform, for example, was an appropriate 'market size' (World Bank 2004b; Vedavalli 2007). According to World Bank publications, a large market size and a high density of current or potential electricity consumers would create an incentive for many private operators to function simultaneously. The Bank also noted the importance of a mature, well-developed set of network facilities, that is, a sound infrastructure network. The presence of good infrastructure facilities would reduce the complexity of private firms providing services as the potential incentive problems associated with negotiating both service requirements and infrastructure investments would be avoided. In addition, the political capacity and support to execute reforms, and sound institutions capable of managing and implementing reform, were also noted as valuable conditions, highlighting the role of the government's political commitment to 'effective policy implementation' (Vedavalli 2007, pp. 330–5).

On one level, the decision to reform and/or restructure an electricity delivery system is straightforward. While various options exist for the structure and organization of an electricity market, one of the chief requirements of reformers is to decide what *process* will be followed to reach the desired end point: 'successful policy outcomes depend not simply upon designing good policies but upon managing their implementation' (Brinkheroff & Crosby 2002, p. 6).

For utility sector reform generally, it is well recognized that unbundling and privatization can be difficult political actions owing to public discontent with reform outcomes, most notably relating to increases in price, and because of bureaucratic resistance. But policy change in developing countries is also often difficult because the stimulus for change often comes from sources outside government or from technocrats; the resources needed to carry out change either do not exist or must be reallocated; change requires that government organizations adapt and modify to new tasks; and because change can also be very complex (Brinkheroff & Crosby, 2002, pp. 18–21).

Owing to these difficulties, a continuum of implementation tasks to facilitate success have been promoted, which include legitimization, constituency-building, resource accumulation and mobilization, and organizational design or modification: decision-makers must assert that the proposed policy is necessary and vital (legitimization); a constituency of interests must see the value of the policy change and play a part in marshalling support for that change (constituency-building); to build a constituency of support, the new policy or policy change 'must be of sufficient importance to overcome or at least neutralize the forces opposing implementation' (Brinkerhoff & Crosby 2002, p. 27). A new policy requires that sufficient human, technical, material and financial resources be allocated to see the change or policy through (resource accumulation) in an appropriate manner so that progress is obvious and success can be communicated to constituents (ibid, p. 30); and organizations must change or be modified to respond to the new policy, which is often a very difficult task in face of bureaucratic resistance to new models, and the historic ways that management and responsibility have been organized.

Moving from the general to the specific, in the 1990s and early 2000s, additional implementation and design conditions specific to electricity were also discussed, which were more attentive to environmental and social concerns: (1) intergovernmental, inter-institutional and inter-organizational cooperation and partnerships (ESMAP 2000, p. 105; Ostrom et al. 1993; UNCHS 2001, pp. 142–6); (2) public input in decision-making and consideration in service delivery decisions (Davidson & Karakezi 1993; Karakezi & Mutiso 2000; Mackenzie & Christensen 1993; McGranahan & Satterthwaite 2000; Mugyenzi 2000; see Turkson & Wohlgemuth, 2001); and (3) human health, environment, equity and gender, particularly owing to the reliance on fuelwood and charcoal for cooking and the disproportionate health burden borne by women and children as a result of fuel choice (Davidson & Sokona 2001; Fiil-Flynn and SECC 2001; Hardoy et al. 2001; McGranahan & Satterthwaite 2000; UNCHS 2001; World Bank 1993).

It is notable that the emphasis on civil society participation and environmental and social concerns here is consistent with other analyses of energy-related decision-making processes at the time. For example, in its two-and-a-half-year independent review of global dam construction practices, the first strategic priority the World Commission on Dams recommended when constructing large dams is 'gaining public acceptance' through decision-making processes that 'enable informed participation by all groups of people' (WCD 2000, p. 215).[1] The World

[1] The World Commission of Dams was established in May 1998. Owing to strong criticism about its dam construction practice, in 1994 the World Bank's Operations Evaluation Department (OED) announced it would review the large dams it had funded. The report it produced in 1996 suggested that 74% of the large dams it had funded were 'acceptable or potentially acceptable'. This finding was critiqued by many international NGOs, in particular the International Rivers Network, now named International Rivers. To announce its findings, the World Bank arranged to co-host a workshop with the World Conservation Union (IUCN). IRN, however, received a leaked copy of the report, and in its review argued that the OED had 'wildly exaggerated the benefits of the dams under review, underplayed their impacts, and displayed deep ignorance of the social and ecological effects of dams' (McCully 2001, p. xx). As a result, IRN and other organizations demanded that an independent international review of large dams occur. Hence, from the meeting in Gland, Switzerland, originally intended to share the OED's findings, an agreement was reached that dam builders and their critics would 'work together to review the development effectiveness of large dams and to establish

Bank reluctantly supported the WCD's mandate, but it remained that the Bank's publications on energy and public sector reform placed most emphasis on regulatory and technical issues. Citing different cases, researchers categorically stated that the Bank demonstrated a 'blind trust in privatization' (Pineau 2002, p. 1011), allowed little public input (Karekezi & Mutiso 2001) and neglected environmental concerns and renewable energy technologies (AFREPREN/FWD 2005). Despite this evidence, scholars also note that in the early 1990s the World Bank stated that reform should be a gradual process with the pace dependent on the sector's capability to manage reform: there is a recognized need to 'slow the pace ... out of concern that it is extremely politically difficult to change reform structure or rules after the process is underway ... Nevertheless, these [sic] cautionary advice is not reflected in practice' (Wamukonya 2003, p. 1282). Referring to reforms undertaken in Mauritania, Zimbabwe and Lesotho, which took between two to four years, Wamukonya stated that the project plans the Bank prepared ran contrary to its 'gradual reform' policy position. In contrast, industrialized countries such as the UK, Australia, Spain and Chile used a much slower pace of reform, ranging from eight to ten years (2003, pp. 1282–3). Further, Gratwick and Eberhard (2008) note that: 'By 1999, power-sector reform had gained traction, and a set of measures that became known as the "standard prescription" or the "standard model" was being widely advocated' (in Kapika & Eberhard 2013, p. 5). While this standard model was prescribed and often accepted by countries to secure financial support, the pace and extent of reform was not uniform. Uganda, for example, has implemented more of the standard model than any other country in Africa (Kapika & Eberhard 2013), while its neighbours, Tanzania and Kenya, have proceeded slowly and differently with state utilities continuing to play central roles in their sectors.

   Together, this evidence reinforces the view that in Africa the World Bank was promoting the achievement of technical reform outcomes using a standard model of reform, with much less attention to the process of achieving those outcomes and to the domestic political or social implications of such an outcome. Evidence also suggests that the IMF approached the privatization of utilities in the same way. Citing

(contd) internationally accepted standards that would improve the assessment, planning, building, operating and financing of these projects' (ibid). A Reference Group was created to oversee the establishment of the review. The Reference Group reached agreement that Kader Asmal, South Africa's water minister, would chair the Commission, but over the next few months there was concern that the Commission would collapse owing to disagreement over who else would sit as commissioners. Agreement between dam industry representatives and dam-affected people was eventually reached, and in February 1998, the World Commission on Dams was launched (see Conca 2006; McCully 2001; WCD 2000; and www.dams.org for more details). The recommendations produced by the 12-member international Commission were the outcome of regional consultations that included 1,400 individuals from 59 countries, 947 submissions from 80 countries, 17 Thematic Reviews and 100 commissioned and peer-reviewed papers. Such notables as Professor José Goldemberg and Medha Patkar (Struggle to Save the Narmada River) were members of the Commission.

Côte d'Ivoire's experience in the privatization of its telephone service, Joseph E. Stiglitz writes: 'the telephone company was privatized, as is so often the case, before either an adequate regulatory or competition framework was put into place' (2003, p. 56). Stiglitz goes on to comment on the IMF's rationale for privatizing quickly:

> The IMF argues that it is far more important to privatize quickly; one can deal with the issues of competition and regulation later ... There is a natural reason why the IMF has been less concerned about competition and regulation than it might have been. Privatizing an unregulated monopoly can yield more revenue to the government, and the IMF focuses far more on macro-economic issues, such as the size of the government's deficit, than on structural issues, such as the efficiency and competitiveness of the industry. (2003, p. 56)

Indeed, the political challenges emphasized in the policy change literature, such as legitimization and constituency-building, garnered weak attention in electricity sector reform in many African countries. This raises important questions about the rationale for this approach and the sensitivity of reform proponents to the difficulties its approach produces.

To simply say that the World Bank did not or does not pay adequate attention to the political process and the social and political implications of its reform agenda does not accurately characterize the complex domestic situation within which the Bank operates. Here we see a complex and significant dilemma arising over how reform and development projects should be undertaken. Through public sector reform strategies that are framed under the banner of 'good governance', international agencies, and the World Bank in particular, were influencing and realigning the political and bureaucratic systems of countries, but through managerial emphases on transparency, accountability, independence and efficiency (Harrison 2001; 2005). At the same time, customary 'development projects' continued to be implemented. These projects, however, were taking place at the same time as other dramatic political changes: complex 'second-generation' reforms were occurring; civil society groups were becoming more adept and competent in policy analysis and at challenging the state's policy agenda; citizens and domestic civil society groups were becoming accustomed to more opportunities to participate in the political system; and governments were frustrated and antagonistic to internationally mandated reform and/or review processes that would slow down desired outcomes. As a result, the complexities of what would seem to be straightforward proposals for reform or project implementation increased by several magnitudes, not least because of the long list of technical and political conditions deemed necessary to execute all or even some of the basic principles or tenets of energy sector reform identified earlier and summarized here:

- Large dense market
- Good quality infrastructure
- Sound reform design
- Strong macro-financial and energy sector linkages
- Political capacity to execute reforms
- Sound institutions capable of managing and implementing reform
- Policy legitimacy
- Constituency-building
- Resource accumulation and mobilization
- Organizational change
- Intergovernmental, inter-institutional and inter-organizational cooperation
- Popular participation in decision-making
- Integration of health, environmental, gender and equity concerns.

This list of 'ideal conditions' is daunting – an almost impossible set of technical and political conditions suggested as needed for reform success. Add to this the expectation of some public involvement in electricity reform – even tokenistic involvement – and the challenge intensifies: while 'participation is not a panacea for implementation success' the general sentiment remains that 'participation is helpful, even essential' as a threshold condition (Brinkheroff & Crosby 2002, p. 52). Despite this, in the name of speed and financial well-being, it remains that technical concerns continued to dominate electricity reform in the 1990s and 2000s. Further, a clear disconnect was emerging between the general policy advice being promoted by the World Bank in Washington, DC and what was actually happening in countries.

Energy reform experiences in Africa that had taken place were acknowledged to have marginally improved the technical performance and viability of power sectors by bridging short-term generation shortfalls and enhancing the financial health of state-owned power utilities (AFREPREN/FWD 2005, p. 117). African energy analysts argued, however, that separating technical conditions from political conditions in future reforms would reinforce ongoing concerns with the long-term sustainability of energy sectors, namely, that public discontent will increase if unable to offer input, further slowing reform; that investments in a diverse range of energy sources, including renewable energy sources, will remain minimal; that rural and urban poor populations will remain without electricity for decades to come; and that electricity will remain costly in the short, medium and long terms (AFREPREN/ FWD 2005).

Assuring that the above conditions transpire in a reform process is a daunting challenge and demands a great deal from whoever is driving the process – government or international donor. As South Africa's 1998 White Paper on Energy Policy cautiously noted, despite the appeal of integrated, multi-interest planning for energy, this goal suffers 'from the same drawbacks as other ideal models, in that it requires an enormous amount of data and analysis to implement' (Republic of South Africa 1998). The report continued: 'For various reasons South Africa has very limited energy data and, furthermore, very limited capacity to perform this sort of policy analysis.' If the observation about limited capacity was made in the context of South Africa, a country generally recognized for having one of the more inclusive and sophisticated national policymaking processes in the sub-continent, it should not be surprising that this was equally or more difficult in other African countries, including Uganda.

Capacity to manage and/or coordinate multi-interest participation is extremely difficult. As Anil Hira, David Huxtable and Alexander Leger bluntly explain in their globally comparative study of citizen participation in electricity regulation: 'Including the public on a large scale is messy' (2005, p. 57). For these authors, participatory processes can require major expense and effort in public education; when processes are done poorly, sometimes only the most interested consumers end up participating (therefore, distorting public input); and, from a political perspective, including the public has the potential to 'expose different factions and ideas', and generally make controversies more intractable (ibid). These sentiments are consistent with general policy analysis, which notes the high potential for problematic outcomes if policy managers do not put considerable time into identifying the goals of participation while at the same time answering questions about how participation will be managed, determining who can participate, and when in the decision-making process participation will occur (see Brinkheroff & Crosby 2002). In short, theory and practice show that reform processes matter substantively. As will be illustrated later, for example, Ugandan MPs were also keenly concerned about the process of reform and the rationale for reforming the electricity sector. In 1998, Benedict Mutyaba, Chairperson of the Ugandan Parliamentary Sessional Committee on Natural Resources, inquired why a law amending the electricity sector had been introduced before a policy providing national direction to the sector – a key principle articulated by the World Bank. He stated:

> ... the Committee, like I would think all MPs, supports the policy of liberalisation of power generation ... [but] you do not start with the law and go to the policy, you start with the policy, and Members, we have not seen the policy. I do not know whether there is any Member here who has seen the policy on power generation. (Hansard, Government of Uganda, Thursday, 27 August 1998, p. 4732)

Other problematic public engagement processes were also dominant in Uganda, as Chapter 4 reveals. With respect to the Bujagali dam, for example, public meetings about the project were often raucous events, with little clarity about whether attendees were participating on their own volition or had been paid to attend.

Given that research suggests that the process of decision-making and reform is important, and even more so when engaged in complex reforms, what is the record of technical, social and political conditions or considerations intersecting in reform debates? The next section returns to the case of Uganda to illustrate how the general principles espoused for sound reform became embedded and taken up in the country. The section returns to the mid-1980s when the World Bank and the new national government of President Yoweri Museveni started to focus on reform and privatization following national political and macroeconomic stability. This early post-conflict period (post-1986) reveals how technical improvements in the economy established an important precedent for how electricity and other reforms would evolve in the 1990s. It also reveals the character of energy governance in Uganda that would emerge in the 1990s, and how the broader political context began to shape how donors and the national government would respond to its looming electricity crisis.

How Uganda's electricity reform model emerged

When Yoweri Museveni and the National Resistance Movement (NRM) came to power in 1986, they:

> inherited a country whose economy was in ruin and whose political and administrative institutions were in tatters: the infrastructure was dilapidated ... social services particularly health and education, were in a sorry state; inflation was running at over 120 per cent annually; government was running a large budget deficit due to financial indiscipline; the country had a serious balance of payments problem, which was compounded by over-reliance on coffee exports for foreign exchange earnings; the exchange rate was overvalued by almost ten times the market rate; and income per capita was only 59 per cent of its 1971 level. (Kiyaga-Nsubuga 2004, p. 89)

In response, in 1987 the government announced its Economic Recovery Programme. External support for the programme marked the beginning of President Museveni's long relationship with the World Bank and International Monetary Fund under the auspices of a 'series of consecutive and sometimes overlapping structural adjustment loans' (Dijkstra & Kees van Donge 2001, p. 842).

The success of Uganda's macroeconomic reforms in the early to mid-1990s are well documented, but as macroeconomic stability emerged, so too did concerns about the growing authority of the President, the weak role of formal institutions and civil society organizations, and the influence of multilateral agencies (see Tripp 2004; 2010, for example). In particular, many wondered why donors were less critical of Uganda's no-party political system (Dijkstra & Kees van Donge

2001; Harrison 2001; Muhumuza 2002), thus holding it 'to a different standard than many other African states' (Tripp 2004, p. 19) and 'leaving the momentum for political reform in the hands of Ugandans' (ibid, p. 23). Graham Harrison noted that Uganda was only criticized mildly for its no-party democracy, even though the general political direction that it was taking was towards a single-party state (see Hickey 2005): 'The situation in Uganda can be compared with the conditionality politics of its neighbour Kenya, where the governance agenda is constantly reinforced and invigilated through the threat or execution of the withdrawal of external funding' (Harrison 2001, p. 659). Harrison went on to say that this is '[m]uch to the chagrin of Kenyans, who find it unjust that Kenya has gone through the pains of transition to multipartyism but is still less favoured than the "no-party" state of Uganda' (p. 659). In the 2000s donors have been more willing to censure and withhold support to Uganda for policy measures relating to such things as grand corruption, military spending and anti-homosexual laws. In the early 2000s, donors did more frequently and more openly express disquiet at the slow nature of progress towards democratic governance in Uganda (De Coninck 2004, p. 61), but never initiated the same degree of censure as in Kenya in the early 2000s, for example.

Of all bilateral and multilateral donors operating in Uganda, one figures most prominently – the World Bank. In 2003, popular Ugandan journalist and political commentator Andrew Mwenda described the relationship between the Bank and Uganda as a 'marriage of convenience': 'The World Bank needs Uganda as much as Uganda needs the World Bank' (Mwenda, interview, 17 January 2003). Similarly, a bilateral donor representative with whom I discussed the World Bank's role in electricity sector reforms confidentially, yet light-heartedly, remarked: 'People refer to Uganda as the Pearl of Africa, but some people say that Uganda is the Pearl of the World Bank' (Anonymous interview, European donor representative, 18 March 2002). Others have argued that Uganda has served as a 'fertile ground on which test new approaches' relating to poverty alleviation and economic development (De Coninck 2004, p. 60).

Why are these perspectives important to consider in the context of electricity? Given Uganda's early success with macroeconomic reforms and donor support in these efforts, it is important to ask why electricity appears to be such a different and difficult policy sector to ameliorate. This is particularly important considering the important relationship between economic performance and electricity sector performance. One of the interesting characteristics of electricity is that the sector's performance can both undermine and be undermined by the economy. In the latter case, Ranganathan writes, 'one of the key bottlenecks preventing the development of the power sector turns out to be the economy' (1998, p. 5). Given that electricity cannot be stored, if demand decreases owing to poor economic conditions, then a power company is paying for electricity generation that is not consumed. Leaving aside potential losses due to theft and technical problems, financial losses can be compounded by the devaluation of a currency in a weak economy. But the more common relationship between energy and economy is the reverse: poor energy supply undermines economic productivity.

If firms and households in Uganda were seeking better electricity service amidst a growing economy, and early public sector and economic reforms were generally held to be successful, why was electricity sector reform so difficult? Part of the answer lay in how the relationship between donors, the state, and non-state actors evolved in Uganda – how the character of governance changed over time and at the most critical early moments in reform.

Privatizing Uganda

When the National Resistance Movement first took power, Museveni initially took a stand against international financial institutions (IFIs) (Kjær 2004a, p. 396), stressing 'their complicity with the Obote II regime and their role as agents for external intervention' (Harrison 2001, p. 662). Imposing policies running counter to liberal-economic orthodoxy – price and foreign exchange controls to curb inflation – enforced his position (see Kjær 2004a). These initiatives, however, did not produce the desired results and inflation accelerated. Hence, the decision was made to move the economy towards a market-liberal framework, with some believing that in exchange for donors accepting Museveni's 'no-party' political system, Uganda would permit, among other things, a dramatic reduction in the state's economic controls. In a 2005 editorial, Andrew Mwenda expressed this hypothesis:

> Museveni needed money as a political resource to consolidate power; Uganda needed stability and economic recovery, donors needed a country where they could pursue economic policies favourable to international capital, but which they also thought they could produce an 'economic success story' in an otherwise distressing African continent ... In other words, what was good for Museveni to consolidate his power was coincidentally good for Uganda, but also good for the donors who were searching for an African success story. (Mwenda 2005)

The result? 'Museveni gave donors almost complete control of the economic policymaking process, and in return the donors allowed him a free hand to pursue his preferred political and security machinations like banning political party activities in the country and pursuit of military adventures at home and in the region' (Mwenda 2005). Not surprisingly, Museveni would passionately disagree with this sentiment. Quoting from his 1997 autobiography, *Sowing the Mustard Seed*,

Kjær notes Museveni's position: 'We did not adopt market economics as a consequence of pressure, but because we were convinced it was the correct thing to do for our country. If we had not been convinced, we would not have accepted it' (Kjær 2004a, p. 396). Others also commented on Museveni's decision to reject his earlier intellectual leanings towards socialism (Hyden 2006), arguing that what is clear is that Museveni 'initially conceded considerable autonomy to economists and techno-crats who were free to design neoliberal economic policies that reduced state involvement and encouraged private investments' (2006, p. 131). Thus, Museveni wanted to 'make a break from the past' (Hyden 2006, p. 132).

The second anniversary of the National Resistance Movement-led government fell on 26 January 1988. Celebrations took place throughout the capital city, Kampala. In the words of Mahmood Mamdani, these celebrations were to mark peace and security, but not economic well-being (Mamdani 1988, p. 1155). Museveni's initial Economic Recovery Programme had not reduced inflation, and, in Mamdani's eyes, was no different than Amin's 'Action Programme' or Obote's 'Reha-bilitation Programme' (Mamdani 1988, p. 1163). The World Bank had supported the government's early reform efforts with its first Economic Recovery Credit (1987–89), but despite continued economic growth the poor stabilization results produced an evaluation of 'unsatisfactory' (Dijkstra & Kees van Donge 2001, p. 843). Thus, by 1989, a second credit was provided (1990–92), but produced little better result in the eyes of the Bank.

Up until 1992 the relationship between Uganda and donors had remained uneasy; macroeconomic stabilization had proven elusive despite increased and leniently provided aid (Dijkstra & Kees van Donge 2001, pp. 842–3). During the period 1987–92, donor demands for privat-ization of parastatals and further devaluations were debated within Uganda but resisted (2001, p. 843). By 1991, donor frustration was esca-lating, and came to a head when Uganda refused to address concerns surrounding foreign currency and exchange rate controls. This culmi-nated in some donors severing aid (Djikstra & Kees van Donge 2001). Despite reduced income, the Ministry of Finance did not cut expendi-tures, producing a budget deficit leading to an upsurge in inflation. Making matters worse, the IMF suspended programme aid, and other donors demanded 'firm control of government expenditure and full liberalization of the exchange rate' (2001, p. 843). At this point, only two realistic options were open to Uganda: accept donor advice quickly, or accept it less quickly (Hansen & Twaddle 1998, p. 7). Museveni chose quickly.

Museveni reshuffled his cabinet, 'removing a hostile finance minister under advice from the IMF' (Harrison 2001, p. 663) and appointed a proponent of fiscal discipline, Emmanuel Tumusiime-Mutebile, as Permanent Secretary (PS) (Dijkstra & Kees van Donge 2001, p. 843).

Tumusiime-Mutebile had already been the PS of the Ministry of Planning and Economic Development. The two ministries were merged to create the Ministry of Finance, Planning and Economic Development (MFPED). Under the new PS, a cash budget was introduced so that expenditures did not exceed the sum of revenues and foreign grants and loans, and the relationship between donors and the government grew more cordial (ibid). From 1992 onwards Uganda's economic stability and success started to materialize, although not often as clearly as is suggested in the mid-2000s (see Hickey 2005). Aside from the President, the Ministry of Finance became the dominant institution from which economic and public sector reform and privatization would follow.

In Africa generally, and Uganda specifically, the role and significance of the Ministry of Finance is unparalleled. The Ministry of Finance 'serves as a conduit between the state and the donor/creditors. In both Tanzania and Uganda, all agreements for project and programme funding are signed with the Permanent Secretary of the Ministry of Finance, regardless of the "target" ministry' (Harrison 2001, 664). Indeed, the Ministry of Finance in Uganda had a disproportionate degree of power in the late 1990s:

> ... the Ministry of Finance has received a disproportionate amount of training and technical assistance, that is externally-funded posts for experts (almost always expatriates). Donor assistance allowed MoFPED to establish an Economics Masters degree course at Makerere University, taking in twenty employees per year; donors have funded research groups within MoFPED with a view to improving the technical competence of economic planning and policymaking; and the World Bank and UNDP have introduced incentive schemes into MoFPED to enhance performance and motivation ... Within other ministries, the perceived expertise within the Ministry of Finance gives it an image of power which is reinforced by the larger and better maintained premises and the high level of computerization of the ministry. (Harrison 2001, p. 665)

As the lead bureaucrat, the Permanent Secretary in East Africa is also unequalled.[2] The Permanent Secretary of Uganda's Ministry of Finance during the 1990s is well known: a Uganda-based USAID official interviewed by Harrison made this remark about Tumusiime-Mutebile [the PS]: 'When he left Kampala, all the donors panicked because all of their projects went through him' (2001, p. 665). Adding to the power of the Ministry of Finance, the accounting officers of all other ministries (those who control the budgets) 'are centrally appointed by the Secretary of the Treasury [also the PS of Finance]. This makes all ministries constantly aware of the power of the Ministry of Finance and its central

---

[2] In Uganda, Kenya and Tanzania, the chief bureaucrat for each Ministry is generally titled the 'Permanent Secretary'. Elsewhere in Africa, this position is similarly titled 'Director General', 'Chief Executive Officer' (CEO) or chief 'Accounting Officer'.

concern – fiscal prudence' (Harrison 2001, p. 666). Hence, along with donor support, the President, the presidentially appointed permanent secretaries, the Ministry of Finance and MFPED-appointed accounting officers represented a close-knit group of actors guiding and owning early economic reforms. Macroeconomic reform did not exist in isolation, however. Museveni's early commitment to reducing the size of the public sector along with the privatization of parastatal companies was also central to Uganda's early economic strategies (Kjær 2004a, p. 397). The dominant role of the Ministry of Finance, donors and the President in macroeconomic and sector reform was felt directly in other sectors, which is where the electricity sector emerges. One challenge lay in whether the capacity existed to implement complex reforms at the sector level.

In 1989, alongside the establishment of the Presidential Economic Council, President Museveni appointed a Public Service Review and Reorganisation Commission (PSRRC). After meeting initial resistance within the Ministry of Public Service, as with the Ministry of Finance, Museveni took personal control and reduced the number of ministries, appointed a new Minister of Public Service, and appointed an 'Implementation and Monitoring Board' (Kjær 2004a).

Explaining Museveni's role in reforms, Kjær quotes a member of the reform commission:

> In a surprising and swift stroke of the pen, President Museveni reduced the numbers of ministries from 32 to 21 in 1992 ... This action by the president was a strong message to the conservative mainstream of the civil service. If ministers could be removed from office to promote efficiency and economy, then similar retrenchment had to be carried out at all lower levels of the government. It opened the gate. 11 out of 32 permanent secretaries were removed. (Katorobo 1996 in Kjær 2004a, p. 397)

Initially, public sector reform progressed quickly, with 150,000 retrenchments, large-scale pay reform and major financial and functional decentralization measures: 'The civil service reform programme was clearly perceived as part of a wider rebuilding project which the NRM government had undertaken, consisting of structural adjustment, decentralisation, constitutional and electoral reform ... a rebuilding project which was the whole *raison d'être* of the NRM regime' (Kjær 2004a, p. 397). Hence, a central part of structural adjustment in Uganda was PSR and state divestiture from public enterprises (PEs) or parastatals – privatization.

In the late 1980s, donors had been pressuring for the privatization of PEs in Uganda (Dijkstra & Kees van Donge 2001; Tangri & Mwenda 2001; Tukhabewa 1998). Initial resistance to donor arguments for privatization subsided in the early 1990s. At this time, Uganda had 156 PEs, many operating at a loss, with low productivity, and were characteristic of PEs in developing countries – they performed poorly (Nellis & Kikeri 1989; Tukhabewa 1998).[3] Privatization, therefore, became an important component of structural adjustment in Uganda and was formally introduced by the NRM in 1991 under the auspices of a policy for public enterprise reform and divestiture (PERD), and under the guidance of the newly formed Public Enterprises Reform and Divestiture Secretariat (PERDS). In 1992, a list of 40 PEs to be divested was released, including large parastatals relating to banking, insurance, railways and telecommunications (Tangri & Mwenda 2001, p. 118). All parastatals were classified under five categories – retain, majority share, minority share, fully divest, liquidate. In 1993, the Uganda Electricity Board (UEB) was classified a Class 1 enterprise, *to be retained*. Maintaining electricity as a public monopoly was consistent with the history of electricity expansion globally, but also a recognition of the value and importance of the utility domestically. Yet only a few years later a law would be introduced to unbundle the UEB and reverse the initial goal to retain the electric utility as a state monopoly.

The initial round of privatization and divestiture in Uganda began prior to the passage of legislation supporting it. Hence, while the initial process was consistent with a linear model of policymaking – identify problem, develop a policy outlining intended actions, create necessary regulation or legislation to execute policy – the final stage of this model, implementation, began prior to the necessary legal framework being in place to execute the policy. This tendency repeated itself in other sectors, like electricity and forestry, and has since emerged in relation to other complex sectoral reforms such as biotechnology (see Schnurr & Gore 2015). As a result, several issues and concerns arose in the early period of privatization, particularly surrounding the rationale for privatization, along with transparency, corruption and ownership of reforms.

Critics of Uganda's early privatization process suggested that 'little was done to educate the public about the policy of privatization and its potential benefits' (Tangri & Mwenda 2001, p. 118). The lack of a communication strategy and citizen participation in the privatization policy in fact led Parliament to suspend the sale of PEs in early 1993 (Tukhabewa 1998). It was only after a closed session of Parliament that the law passed (1998, p. 65): the national government belatedly 'embarked on a propa-

---

[3] This is a broad generalization and there are gradations of performance historically. Nellis and Kikeri (1989, p. 659) note that evidence from the early 1980s shows that PE sectors in 13 African countries accounted on average for 17% of GDP. In Latin America and Asia, where evidence is sketchy, GDP contribution was sometimes at 17% or below. Meanwhile, in some African countries, in the 1980s PEs accounted for upwards of 40 to 60% of GDP. Examples here include Algeria, Egypt and Zambia. Nonetheless, the authors characterized the performance of PEs in the 1980s in the following manner: 'too many PEs cost rather than make money; and too many operate at low levels of efficiency' (1989, p. 660). Asian countries showed fewer problems than Latin America, which in turn performed better than PEs in sub-Saharan Africa.

ganda campaign through advertisements in the newspapers, radio and a drama group' to persuade the public that privatization has been and will be beneficial, producing better jobs, education and health (1998, p. 65). Despite the initial push, privatization of state-owned enterprises was hardly implemented until 1995, and was then carried out slowly (Dijkstra & Kees van Donge 2001, p. 843). Part of the difficulty was that reform leadership remained confined to a very small number of people: placing faith in the ability of the President and permanent secretaries to own and drive reforms created a potentially 'fragile' scenario whereby a small number of individuals own the knowledge for change (Dijkstra & Kees van Donge 2001, p. 845). It also created a precedent for reform and decision-making and governance generally: the spaces available for debate, the actors involved and the knowledge used to deliberate were limited or restricted. Ironically, delays in the privatization process and concerns with corruption led donors to ask President Museveni to take personal charge of privatization in 1998 (Dijkstra & Kees van Donge 2001, p. 843), while others concurrently expressed concern with the lack of transparency and poor communication of reform benefits to citizens (Tukhabewa 1998, p. 65).

Key bureaucrats leading public sector reform acknowledged the difficult time elected officials and citizens had explaining and understanding the rationale for privatization. Members of Parliament (MPs) were not and had not been very good at communicating the rationale for privatization; those most knowledgeable about reforms, public servants, were restricted from speaking out publicly; and the public perceived that government got a bad deal from the sale of state-owned enterprises and that corruption was rife (Interview, Emmanuel Nyirinkindi, Director, Utility Reform Unit, Ministry of Finance, 14 May 2002). These observations are consistent with other national experiences with privatization (see Birdsall & Nellis 2003). On the point of revenue from the sale of public firms, part of the problem in Uganda was that the physical assets of corporations were valued at more than the public's perceived market value of an enterprise. The Ugandan public had a 'very high but very unrealistic expectations of what was achievable when selling public enterprises', particularly because 90% of Ugandan public enterprises were carrying massive debts (Interview, Emmanuel Nyirinkindi, Director, Utility Reform Unit, Uganda, 14 May 2002). As a result, the Director of Uganda's Utility Reform Unit at the time suggested that the government incurred strong criticism when the public didn't understand the chief purpose of selling – relieving the government of the financial burden of the enterprise. Further, when comparing Uganda's divestiture process to other sub-Saharan African countries in relation to total number of sales and value gained from sales, the country had done well. Hence, from the perspective of the Ministry of Finance, while the privatization process has had difficulties, one of the central issues confounding the process was a problem of communicating the intent of the process to both citizens and elected officials – a problem undermined by the unwillingness of elected officials to communicate the rationale for reform publicly.

Privatization and public sector reform remained priorities in Uganda in the early 2000s. In fact, as reforms to the electricity sector were gaining momentum, the Ministry of Finance was planning to turn to the National Water and Sewerage Corporation next, preparing it for privatization. However, it was monitoring the experience with electricity and being 'very cautious' as a result of the challenges with the electricity sector (Interview, David Ssebabi, Team Leader, Utility Reform Unit, 8 January 2003). It is noteworthy that the privatization of the water corporation never came to be and became celebrated as an example of the successful reform of a public enterprise providing a key public service – water (Muhairwe 2009). It is also noteworthy, but not surprising, that many in the Uganda Electricity Board also explained in interviews with me that they were not given enough opportunity to reform the sector before its unbundling was announced. The challenge of reforming complex public enterprises was persistent in Uganda. Indeed, early in 2015, it was reported that President Museveni informed his cabinet that they would not privatize any more public institutions (Nakaweesi 2015).

By 2006, the World Bank ceased supporting the Privatisation Unit, although it continued to operate. While the Bank publicly stated that it is was no longer supporting the Privatisation Unit because programme funding for the unit had lapsed, media suggested that political interference, lack of transparency and the slow pace of divestiture were key reasons for the World Bank's decision (*East African*, 14 March 2006). With respect to political interference, several events were cited. In March 2006, President Museveni publicly proclaimed that the Privatisation Unit should award the concession of the Kinyara Sugar Works to a domestic firm rather than an international one (ibid). This followed another incident where the President publicly proclaimed that the Dairy Corporation Ltd should be sold to a Thai company. Thus, the President was seen as playing a dominant role in divestment decisions if not interfering in the unit's independence. This criticism or trend also arose with the electricity sector, as will be soon noted. In the meantime, public sector reform remained central priorities in the country. In 2005, the World Bank committed another $150 million to Uganda's Poverty Reduction Support Credit (PRSC4) – its fourth. Uganda's third five-year World Bank-funded Public Sector Reform Programme (2002–07) also made public sector reform a priority, particularly in light of the fact that the public sector deficit had risen from 6% of GDP in 1997–98 to over 12% in 2001–02 (MFPED 2004). But while public sector reform remained prominent, public servants outside of the Ministry of Finance were unclear of the model of reform being chosen and the benefits.

Concerns with privatization and reform agenda

One of the consistent ways that sector reforms have been undertaken in Uganda is to move service delivery and regulatory functions outside government, usually under regulatory authorities or semi-autonomous agencies. This approach, the 'executive agency model' (Therkildsen 2000), was intended to leave government with policymaking responsibilities while revenue-generating and service provision activities were to be done at arm's length from government. According to a senior anonymous source, there was 'not much known about how they [independent authorities] are doing in Uganda' (Interview, senior civil servant, Ministry of Public Service, 23 May 2002). Using the example of the Uganda Revenue Authority (URA), a senior civil servant with the Ministry of Public Service explained that the revenue the URA had collected had decreased while its operating costs had increased, registering a 5 billion shilling loss for one financial quarter (*Monitor* 2006b).

In addition, despite the intent of separating policy functions from service delivery and management functions, other Ugandan regulatory authorities in the early 2000s, like the National Environmental Management Authority (NEMA), were given specific responsibility for policy development – something outside their mandate (Interview, Senior NEMA civil servant A, 28 February 2002; Senior NEMA civil servant B, 5 March 2002). As a result, there were functional overlaps and problems of accountability and responsibility, as roles were not clear or consistent. In the case of NEMA, the independent authority was seen as 'a bit schizophrenic' because it crossed so many roles (Interview, European bilateral donor, 28 May 2002). This observation was reinforced by the fact that at the district level, District Environment Officers were supposed to be overseeing regulation but were also planting trees, hence, they were seen as a regulating agency but also doing operational work. Overall, then, one assessment of the relationship between divestiture and public sector reform was that 'it was not very well thought through' (Interview, senior civil servant, Ministry of Public Service, 23 May 2002).

These observations resonate with other experiences in East Africa during this time (see Therkildsen 2000): 'It is hard to escape the conclusion that we seem to know more about what does not work in the public sector in a poor donor-rich country ... than we know about how to improve and sustain performance under present and foreseeable economic, administrative and political conditions' (2000, p. 70). Therkildsen observed that in the 1980s, Tanzania suffered from 'projectitis', when more than 2,000 development projects – mostly donor-funded – appeared in the budget (2000, p. 62). By the early 2000s, Therkildsen argued that 'reformitis' had emerged; that is, a multitude of mostly donor-funded reforms are implemented or under preparation at the same time. In Uganda, a similar situation existed.

As an example of the high administrative burden World Bank projects placed on the Ugandan state, in the period 1991/92–1993/94 the World Bank required 86 policy reforms in the country as part of its conditional lending (Harrison 2001, p. 668). Between the years 2000 and 2005 the World Bank was financing 57 different projects. This number included projects that were approved during the 2000–05 period, or that would close during this period, hence, projects that the Government of Uganda was administering. Apart from three projects that were grants, the remaining 54 were IDA loans totalling just under US $3 billion. When considering these numbers, it is important to note that they do not factor in the range of other bilateral agencies with projects in Uganda, which would include Denmark, Norway, Sweden, the UK, the United States and Germany, to name a few. As Therkildsen noted with respect to Tanzania, there is a strong propensity for donors to support many reforms at the same time.

The emergence of a high number of simultaneous reforms presented serious challenges to small government agencies in Uganda, particularly the Ministry of Energy and Minerals Development (MEMD). Based on his research, Therkildsen made several other observations worth listing: (1) reform seems to breed further reform; (2) donors are directly involved in all public sector reforms, yet it is difficult to ascertain the precise or direct influence of donors; (3) it is difficult to identify strong domestic support for reform; (4) there is no clear evidence of service improvements following reform, particularly under an 'executive agency model'; and (5) it is mainly political–administrative elites that seek to influence the reform process and not various interest groups (Therkildsen 2000).

In the early 2000s generally and specifically for the electricity sector in Uganda, these observations held true: donors were deeply invested in reforms, but from the outside there was little clarity on who exactly was driving different components of reform or programme development; reform decisions were led by a small group of elites; there was little public discussion or opportunity or space for discussion about privatization or energy investments. The outcome of this context, as will be highlighted in Chapter 4, was a complex, challenging and problematic process, which led the Government to abandon the policy and programme implementation methods it relied on in the early 2000s, and to turn towards new development partners, particularly China, in what might be considered a third era of electricity reform: the first era was 1986 to 1996, when donor and government support for the public electricity monopoly continued; 1996 to 2006, when the government unbundled the electricity sector and put faith in the private sector to build dams to ease the supply shortage; and 2006 onwards, when the Bujagali project was resurrected and the government began to seriously explore alternative partners and models for electricity sector improvement.

In the early reform years, Uganda's economic success resulted in it being viewed as a 'good performer' (Dijkstra & Kees van Donge 2001,

p. 841). Along with this observation came the recognition that, unlike in other countries, in Uganda 'donors ... evaluated *outcomes* rather than *processes*' (ibid, emphasis in original). And even if the individual role of donors was not clear to outsiders, their imprint on the process was prominent: with donor funding came 'a new set of regulations concerning the technique of the policy process' (Harrison 2001, p. 670); 'Donors [did] not just impose conditionalities; they also work[ed] in routinized fashion at the centre of policymaking. Donor-funded technical assistance ... [introduced] new methodologies of policy design' (ibid, p. 671).

Before turning to Uganda's specific experience with reform and dam construction in the 2000s, it is important not only to explain the overall context for reform that was emerging from the 1990s to mid-2000s, but also the character of political and policy decision-making as it provides a window into another component of national governance – trends in state–non-state relations – that would also have a meaningful impact on the electricity sector.

The politics of policymaking and pathway choice

Just as Uganda's post-1986 reforms demonstrate an important and unique relationship between the national government and donors, so too does the way policymaking evolved in the country and the way the national government interacted with citizens, civil society groups and donors over time. On one level, the post-1986 Ugandan state was characterized by increased participation of a range of policy actors (Ssewakiryanga 2004, p. 74). On another, in the early 2000s, there was a general observation in East Africa that interest groups rarely tried to influence reform processes (see Therkildsen 2000). On the surface, these two perspectives seem to be contradictory. But the reality was that the precedent for participation without substantive influence was well established by the late 1990s when electricity reform was under serious discussion.

The colonial regime in Uganda initially encouraged domestic civil society organizations and trade associations, as well as their predecessors such as mission-established hospitals and education establishments (De Coninck 2004, pp. 52–3). The colonial state also, however, regulated these organizations, 'forging a symbiotic relationship with civil society' whose characteristics were still much evident in the early 2000s (ibid, p. 52). Shortly after World War II the colonial government's relationship with CSOs changed. Nationwide unions were banned in 1952 following the increased political activism of unions, trade associations and cooperatives. Thus, the colonial period was marked by guarded and regulated openings and cooperation with civil society organizations.

Little changed following independence, with the cooperative movement, for example, expanding at the same time as becoming bureaucratized, and 'distinctions between civil society and business and between civil society and state becoming more blurred' (De Coninck 2004, p. 54). Civil and political conflict and repression, along with poor economic conditions during the Obote I and Amin periods, led civil society organizations to operate narrowly in health and charitable activities. But following their ouster, international NGOs focused on relief in the northeast region of Karamoja and international donor agencies under the guise of Uganda's first structural adjustment programmes grew (ibid, pp. 55–6).

The National Resistance Movement's ascendancy to power in Uganda ushered in a new era for civil society organizations. Alongside the emerging economic and public sector reforms,

> This period of reconstruction provided a space for the emergence of indigenous civil society organisations ... With social service delivery still beyond the capacity of government with donor funding to NGOs in Uganda no longer compromised by political instability, a laissez-faire attitude by government towards NGOs characterised the late 1980s and early 1990s, *so long as they had no political agenda ...* This era of growth for civil society organisations, with most engaged in service delivery, accelerated as the World Bank and other donors forced fiscal orthodoxy upon government. Seen as ideologically preferable to state delivery, CSOs were considered less corrupt and closer to the people. This was the heyday of NGOs. Generously funded, they could act with impunity and without reference to government policies ... This era also established two important dimensions of civil society in Uganda: firstly, the lasting association – even equation – of 'civil society' with NGOs, while its other components, trade unions and co-operatives, were being undermined by structural adjustment, liberalisation and retrenchment; and secondly, the tendency for NGO growth to be driven by the availability of donor funding rather than the need to provide a direct answer to specific locally rooted social or political imperatives. (De Coninck 2004, pp. 57–8, emphasis added)

The early NRM period thus saw a situation where CSOs were by default agents of development and assistance while the national government tried to get its house in order. Donors and international NGOs, eager to assist following years of problems, were happy to support CSOs. Hence, an explicit role for civil society emerged: NGOs and community-based rural organizations were relied upon to supplement government's capacity to implement projects directed at the poor (Twaddle & Hansen 1998 in De Coninck 2004, p. 59). As social and political stability started to emerge, and the state started to reassert itself at the local level, particularly through salaried and bureaucratized local councils (Resistance Councils at this time), national poverty and adjustment policies were introduced that started to redefine the role of the state: the 'bonanza years for NGOs were over. The latitude for their involvement in service delivery was narrowing while donors were reconsidering the funding of such activities through NGOs' (De Coninck 2004, p. 60). One clear signpost of this change was the introduction of the NGO Registration Statute, which required all NGOs to register with the Ministry of National Affairs in 1989 (see Tripp 2000; Dicklitch 2001).

The National Board's membership and regulations suggested a high degree of suspicion surrounding NGOs, but the NRM remained accommodating for the most part, largely owing to its inability to enforce rules and monitor activities. The Board also had weak coordination and planning capacity (see Tripp 2000, pp. 61–2; Dicklitch 2001). As a result, NGO accommodation was a default outcome rather than an expressed position (Tripp 2000, p. 61). Moreover, because of weak internal coordination and planning, 'NGO agendas tended to be donor-driven and Ugandans consequently had little negotiating power' (ibid, p. 62). Despite the early rise in number and influence of NGOs, the de-registration of a handful of NGOs, along with the delay and near denial of registration for others, reinforced the position that 'political' activities would not be accepted (Dicklitch 2001, p. 35). Hence, even though NGOs were tolerated, the NRM was ready and willing to limit their autonomy if there was 'the slightest possibility that they might prove to be too much of a challenge' (Tripp 2000, p. 63).

One of the outcomes of these conditions was that by the late 1990s, Ugandan CSOs, by choice and necessity, worked with the state and donors, rather than challenging their activities and policy choices, with some exceptions like the Uganda Debt Network and Ugandan Law Society. 'The Movement system of government, with its rhetorical and structural focus on inclusion and decentralisation, has subsequently shaped a political landscape where the dividing line between state and non-state actors is blurred' (Brock 2004, p. 95). The reason for reminding readers of this legacy is that it provides important clues about why the early 2000s were a tumultuous period in state–non-state relations, with direct impacts on electricity reform and infrastructure development. Civil society groups began questioning the orthodoxy and choices of government and donors more directly in the late 1990s and early 2000s, and were willing and able to challenge the government in public forums and through formal institutions and structures, such as Parliament and the courts, thus angering the national government, delaying reforms or investments. Thus, the character of state–society relations, the character of governance in Uganda, was changing. The role of donors in this evolution and confusion is significant.

In the mid-1980s and early 1990s donors were promoting and supporting NGOs as important agents of poverty reduction and development activities. While this has not changed, and the number of NGOs in Uganda steadily increased in the mid-2000s (see Barr et al. 2005), concerns with corruption, accountability, poverty and the desire to see a multiparty system led donors to promote a new role for NGOs:

holding government accountable (De Coninck 2004, p. 62). This placed Ugandan CSOs in an extraordinarily difficult situation: donors increasingly wanted NGOs to monitor government activities – even though they were not yet sophisticated enough to monitor compliance and ensure accountability (Interview, Godber Tumushabe, former Executive Director, ACODE, 4 March 2002) – thus encouraging them to be engaged in political activities that were clearly at odds with a government impatient with challenge.

One way that donors tried to mediate this political quandary was to encourage local and national governments to open decision-making and policymaking processes to CSOs, as well as to encourage government to invite CSOs into forums where policies were presented: external actors provided 'opportunities for participation of civil society actors in the policy process, often by encouraging government to create invited spaces for participation. The resultant expansion of spaces for participation was contiguous with a sharp growth in the number of civil society organisations ... [and] ... resulted in a dramatic increase in the range and variety of actors who participate in the policy process' (Brock 2004, p. 95). This became the dominant mechanism for participation in the early 2000s (Brock 2004, p. 103). Two examples help emphasize this point.

In April 2002, I organized a meeting with representatives of fourteen different environmental NGOs working locally and nationally in Uganda. I had organized the meeting to learn how civil society organizations (CSOs) understood and engaged in policymaking processes in Uganda. I began the meeting by asking the important, yet rather awkward, question: 'What is policy in Uganda?' Understanding what is meant by the term 'policy' has important theoretical and methodological weight in policy studies literature, and increasingly in policy analysis in Africa.

In their book *Understanding Environmental Policy Processes*, James Keeley and Ian Scoones explain how the traditional starting point for defining policy is that it 'comprises decisions taken by those with responsibility for a given policy area, and these decisions usually take the form of statements or formal positions on an issue, which are executed by the bureaucracy' (2003, p. 22). 'Conceived in this way, policy is a product of a linear process moving through stages of agenda-setting, decision-making and, finally, implementation' (ibid). The authors, however, continue with the now accepted point that, in practice, policy is 'notoriously difficult to define', and is an inherently political process that does not evolve in a linear manner from a single decision: policies often consist of broad courses of action (or inaction) or 'a web of interrelated decisions that evolve over time during the process of implementation' (ibid).

Citing the work of well-known policy scholars such as Hill (1997), Lindblom (1959), and Kingdon (1984), Keeley and Scoones suggest that the policy process can be characterized in three broad ways. First, policy can be understood in a linear manner that is focused on decisions and the rational behaviour of decision-makers and policy implementers (bureaucrats). Second, policy can be understood as an ongoing course of action that results from bargaining between multiple actors over time – the incrementalist, 'science of muddling through' perspective, where policy entrepreneurs and policy windows resonate. A third perspective is one that is attentive to issues of power – something on which the other two approaches remained silent (Keeley & Scoones 2003, p. 23). In this approach, one consistent with a focus on the character of relations between state and non-state interests – governance – the relationship between knowledge, power and policy are at the centre of analysis.

For one member of the focus group, the answer to 'what is policy in Uganda' was clear: 'The definition of policy in classical politics doesn't apply' (Focus group, 12 April 2002). The participant said that government policy often begins with consultants and is donor-driven. After the policy is drafted CSOs are then invited to participate in a workshop to discuss the policy and comment on it. Hence, in most cases, policy is usually drafted by interests with little knowledge of CSOs, and CSOs are given limited time to comment and reflect on policy proposals. One example given surrounded a workshop meant to discuss a new policy on 'gendering programmes'. The policy document was 250 pages long, the workshop participants had never seen it, and they were supposed to review and comment on it before the end of the day at 4 p.m.

One of the key observations from this experience was that NGOs are typically brought in to the process too late, and must find ways to get involved on their own. I then asked about where ideas for policy usually come from; which actors or knowledge usually drive policy proposals? Not surprisingly, it was unanimously agreed that policy ideas come from donors. But despite this point of agreement, there was not agreement on the extent to which government needed to go to invite CSOs to participate in policy development. One participant remarked that there were 3,500 NGOs in Uganda, so wondered what government was supposed to do – invite all of them? While the group acknowledged that the number of interests made it difficult for government, it should be doing all it can to invite participation in policy development – an observation, however, that is not necessarily consistent nationally. Golooba-Mutebi (2004) for example, adds to this scepticism by providing evidence that popular participation in policy decisions was a panacea owing to time constraints, participation fatigue and fear of repercussions from participating in political decisions.

How did the perceptions of NGOs based in Kampala correspond with the views of Ugandan civil servants? Interviews with several senior bureaucrats in the National Environmental Management Authority tasked with community engagement and public participation in the early 2000s revealed a clear understanding of the tension between the ideal policy and consultation process and reality.

One individual with deep engagement in support to Ugandan Districts suggested that policy can come as a result of three things: a formal process of creating policy; guidelines or a framework for doing something; and, lastly, 'from pronouncements by senior levels of government or people in high positions – when the President says a major focus is poverty eradication then that will be imbedded in and be a central part of policy'. Where does the motivation to change policy come from? A Director in NEMA clearly stated: 'this is usually donor driven' (Interview, 5 March 2002). In this period, policy development was usually part of a more comprehensive package of reforms. For example, a concept paper would be produced by the lead agency for the reengineering of a whole sector, then a sectoral plan would be developed, and then legislation. Several people in NEMA noted that these processes were very consultative. Yet, others acknowledged the weakness in the consultative process and that there was no expectation that consultation would provide any new knowledge or alter the course of reform.

One NEMA policy leader noted that 'Consultation can also serve as a sensitization process and as capacity building; it readies or prepares them [local government or NGOs] to implement' (Interview, 7 March 2002). He noted that consultation should be an effort to help people understand policy, but that the cost of doing consultation can often be quite high and serve as a deterrent. Others in NEMA were much more critical: 'Civil society can easily be hijacked and captured'; when I asked if this individual could think of any NGO in Uganda that had made a meaningful contribution to a decision-making or policymaking process, he replied that he could not think of any: 'They rise and shine and fall – they are not trustworthy' (Interview, 5 March 2002). And what about engaging the general public? Was there any point? 'The public has done very little' and 'usually don't add anything new.' Nonetheless, he noted, there is a need to consult – a need, however, that is greatly influenced by the objectives of the policy: 'The urgency of the issue is also critical; if it is deemed urgent then it will be rushed through,' and the value attributed to consultation will diminish (Interview, 7 March 2002). The complexity associated with the policy also impedes its implementation.

One major challenge for NEMA was how to translate national policy to the local scale. It was noted that, at the District level, there are limited policy formation processes. NEMA had worked with Districts to develop guidelines to try and integrate national policy into local policy, but local government has had a hard time with this. Another problem senior NEMA staff members noted was that in Uganda 'everyone is developing guidelines'. Energy was used as an example to highlight this point. 'Energy touches everything so if there are guidelines for forestry, energy, water, agriculture etc. then it is very difficult; how easy policy is understood affects its implementation.'

Overall, interviews with civil servants and NGOs working on environmental issues in the early 2000s were consistent with respect to the way participation in policymaking was taking place and the motivations for policy development. First, 'participation' was generally understood as a consultation and 'sensitization' process, which NGOs were invited to for them to understand what government was about to do. As a result, NGOs working at the national level had to spend a considerable amount of time staying abreast of opportunities to participate in national discussions. The outcome of this is that CSOs in the capital city

... seem[ed] to be at the level of actively pursuing the single goal of getting their people onto seats in meetings or committees; and reactively responding to any invitation issued to take part in any public forum which might afford profile to the organisation or the issue on which it works. These activities are pursued with apparently little analysis of the impact that they might have: an all-consuming fixation with what might be termed 'the politics of presence' rather than the politics of influence. (Brock 2004, p. 103)

The fact that NGOs were looking to be present at policy forums is not surprising given that the majority of NGO funding was coming from international sources and therefore they were often 'preoccupied with accountability to their donors' (De Coninck 2004, p. 63). In 2001, Barr et al. found that of the 199 NGOs randomly surveyed, more than 90% of the grants received came from international sources, and were subject to a high level of external monitoring (Barr et al. 2005). During this time, NGOs in Uganda were deemed to be quite entrepreneurial, were thought to be led by educated individuals interested in attracting international aid, to be enhancing the well-being of their beneficiaries, and were generally well perceived in the country (Barr et al. 2005, p. 676). At the same time, Ugandan NGOs were also operating as 'subcontractors for international donors' (ibid). Therefore, being present at policy forums provided important opportunities to demonstrate engagement in issues, to promote individual work, and to network and seek out additional opportunities. Hence, for NGOs, the 'politics of presence' was crucial.

Second, when asked how policy evolves, civil servants, NGOs and donor representatives described both an 'ideal' and a 'real' path in Uganda. What is interesting is that those who control the policy process – donors and civil servants – describe a linear path of policy development, although fully recognizing that in practice it never evolves this way. This evidence is consistent with other findings in Uganda at the time; the linear path of policy development existed as a 'necessary fiction' for policymakers and civil servants:

Although patently removed from real life, it [the linear model] is surprisingly alive and well in policy, development and political circles, and even in many policy actors' own accounts of what kind of process they themselves are involved in. The great majority of people we interviewed, when asked, 'What is policy?' gave some version of this linear model. However, when they were asked about the processes that put policies into effect, their descriptions plainly contradicted the linear model. This suggests that the model lives on as a necessary fiction held onto either consciously or subconsciously as a default option, or because few perceive any need to construct alternatives. One reason is that it assigns tangible, definable roles, relatively easily understood and narrated, whereas a model based more closely on real life would be characterised by indistinct roles, blurred boundaries and a high degree of insecurity among most policy actors about the part they play. (McGee 2004, pp. 7–8)

What is so striking about the fictional path of policy is not that it is communicated – it is communicated because it simplifies complex proposals – but because of the frustration that transpires when that linear path is not followed, as is common with controversial or complex proposals requiring the input of many interests and multiple initiatives. This observation is especially significant for electricity in Uganda, given the frustration President Museveni expressed when the Bujagali dam was originally delayed. Again, both civil servants and NGOs clearly recognize that policy does not evolve in a linear manner, but donor representatives and civil servants still describe it this way, and national leaders became frustrated when the prescribed technical path did not materialize.

Third, the motivation for policy reform was clearly identified with bilateral and multilateral agencies. But the degree and opportunity for participation in policy and reform debates, along with the ability to access information on policy issues, was dependent on the issue at stake. One prominent environmental NGO in Uganda working on policy issues, Advocates Coalition for Development and Environment (ACODE), found that information on and participation in social sector policies such as health were much more accessible than those relating to the economic sector (ACODE 2002). Or as De Coninck notes, there are clearly subjects and issues which are not open for debate or for which the national government does not provide or provides only limited invitations, like defence issues (2004, p. 68).

ACODE's research found two other issues of note: (1) sub-sectors that have a higher degree of foreign company involvement or foreign investment like mining and energy are less likely to have opportunities for public participation; and (2) the degree of NGO or donor involvement in a sector or issue correlates with opportunities for participation and access to information. This means that a high number of NGOs working on an issue or sector translates into greater opportunities for participation, and similarly, if an issue or sector receives a lot of donor support, opportunities for participation and access to information also increase.

As Godber Tumushabe, then Executive Director of ACODE, explained: 'Processes for participation are most often because donors make this a pretty explicit requirement' (Interview, 4 March 2002). However, he also suggests that the political costs of participatory processes are not very high because there are minimal fallouts from the types of processes currently seen in Uganda: 'People can participate in processes and never really influence the decisions that are being made.' Moreover, government and donors count consultations, which are better described as 'information sessions', as participatory processes. Despite these concerns, most NGOs and environment-associated civil servants I spoke with routinely pointed to one reform process they considered to be open during this era: the national forest policy and forestry sector plan.

Consistent with Tumushabe's remarks about the relationship between opportunities for participation and civil society and donor involvement, according to Uganda Forest Working Group records, there were approximately 50 CSOs engaged in forestry-related activities in the country (ACODE 2002, p. 35), and the architect and chief financial supporter of the reform process was UK's DFID. Before concluding this chapter, then, it is worth briefly considering how the above general information about policymaking and participation played out in this well-regarded case.

Forestry: A model of reform process success?
In 1998, Uganda's cabinet had decided to divest from the Forest Department and create new legislation, but there was donor pressure to do this systematically (Interview, DFID Representative 1, 28 March 2002). This stemmed from the fact that previous reform processes, such as the Land Act, had not been systematic or comprehensive; immediately following its creation, the Land Act had to be amended. For the forest sector as a whole, the new Ugandan Constitution, introduced in 1995, represented the beginning of change:

> The new constitution brought about widespread reform across government, including proposals for the abolition of the old Forestry Department and creation of the new Ministry of Water, Lands, and Environment (MWLE). The Forestry Department had lost the public trust and was not seen to be carrying out its mandate of policy, regulation, management and services covering all forests. It was also operating under outdated policy and legislation. (DFIDa, n.d.)

Like other reforms in Uganda, the decision was to move forest management and regulation away from government to an independent authority, the National Forest Authority.

According to one independent forestry consultant, the reason for needing this arm's-length arrangement was simple: 'The reason there is a need for a Forest Authority is to get it away from government – there is a strong need to have something independent and immune from political influence; corruption and pay-offs at the local level [were] rampant'

(Interview, independent forestry consultant, 1 March 2002). A DFID official explained the rationale and process of reform were very important given the political sensitivity of the sector. 'The politics of it [forest sector reform] were really important ... there were a lot of fingers in the pie ... [and therefore] needed to get the reform process outside the Forest Department' (Interview, DFID representative 1, 28 March 2002). Hence, in 1999 the National Forest Programme was initiated with multi-donor support (but centrally led by DFID and then GTZ). The initial activity was supposed to be a Forest Sector Review (FSR), which would inform a new policy, law and then plan. However, owing to very significant challenges in compiling accurate data, and what that data might reveal (e.g., more volumes of timber being harvested than was legally permitted), the review took much longer than expected, and in fact, the new policy and plan for the sector were published prior to the completion of the sector's review.

Accordingly, a systematic process would have been to review the sector, then create a policy, then the plan, then create legislation to support the policy and plan. However, there was considerable pressure to 'fast-track' the process and roll it all into one, because it was generally felt that donors would only get 'one shot to pass' the proposed changes. Hence, donors 'had to defy the logic of the process' and develop the legislation and policy in parallel – a fact similar to the early privatization process and something that will be seen for electricity. Another reason for a fast reform process was because Uganda's Public Service Commission and Ministry of Finance set targets for the National Forest Authority to be viable four years after its creation (Interview, NFA Representative, 14 March 2002). In turn, the Forest Secretariat had to take the four-year timeline and work backwards. This was a particularly difficult task given that the decision to create the independent authority was made prior to a sector review, a policy, a plan or legislation (Interview, NFA Representative, 14 March 2002). Again, this deviation from the ideal reform model was consistent with Uganda's early foray into privatization, where the process of identifying and selling parastatals began prior to legislation being in place that conferred legal authority to do so. Nonetheless, in contrast to the criticisms of the early privatization process, the forestry reform process did provide many more opportunities for CSOs to learn about the changes.

The Forestry Secretariat organized nation-wide district meetings in 1999. The consultation process was promoted by the Secretariat, which was led by donors: 'outside donors come in and act as a catalyst' (Interview, DFID Representative 1, 28 March 2002). Interestingly, the opportunities for participation embedded in the reform process were also attributed to the 'participation culture' President Museveni had established in Uganda: 'This is why donors like him – he embraces participation, poverty reduction, and decentralization'; while 'the process [and input from public] didn't add much to the content [of the policy]', former reviews were not open and were opaque. Therefore, while people were sceptical of being consulted, the general sense was that civil society was simply happy to be asked to participate. Initially, the process and draft policies encountered harsh criticism in public meetings, particularly from forest department staff. As a DFID 'learning note' about the process expresses:

> Antagonists used public consultation as an opportunity to denounce the changes in general. The 1ˢᵗ Consultative Conference in particular saw lots of open criticism, even from senior FD [Forest Department] staff at the time. The criticism was of the draft policy and also of the wider changes, including donor priorities … The resistance reflected the general mood of insecurity in the FD and the fact that they felt disempowered and disengaged in the change process in general, particularly with regard to the proposed new National Forest Authority. (DFIDb, n.d.)

Having received public comments on the proposed changes, a draft policy was complete in February 2000 and went forward to the Ministry of Lands, Water and Environment (MWLE). The Ministry reviewed the proposal for four months and made only minor changes. The policy then went to cabinet, where donors supporting the initiative said that bureaucratic inertia further held it up – issues like what colour of paper to use and whether 'secret' should be on every page were some of the issues being debated. By March 2001 the Forest Secretariat received cabinet approval, but then time was spent working out how to launch the policy: 'This process was fraught with politics', namely, 'how to make it high profile.' A year later, in March 2002, the policy was launched at the National Conference Centre in Kampala, an event which I attended.

The title of the conference was the 'Forestry Consultative Conference'. The main conference auditorium was full of hundreds of individuals, and the Minister of Finance, Minister of Environment and several MPs made presentations. For reform supporters, the launch did not matter very much; its purpose was to give public attention to something that many people had been engaged in for some time. The participation and attendance of key Ministers and MPs also provided political weight to the event. Throughout the day, attendees raised questions about the policy that was being presented; however, no substantial discussion took place, and some showed frustration at having no further opportunity to influence the policy outcome. This fact was not lost on government policy actors. A NEMA representative at the launch noted how Uganda often used 'workshops' as another means to create policy (Interview, 7 March 2002). Meetings were usually framed as 'consensus-building' workshops, combining multiple stakeholders and resource users in the case of forestry. However, the setting often 'does not provide equal opportunity for feedback' and does not support politically weak or inexperienced attendees from participating given the intimidating environment.

Since the completion of the forestry reform process, there have been many challenges in the sector. These include mass public protest over the sale of protected lands and ongoing conflict between MPs and decisions made by the National Forest Authority to protect forests, which have called into question the independence of the Authority, just as will be revealed in the case of electricity. Nonetheless, in the early 2000s, Ugandan civil society groups pointed to the regional workshops, working groups and opportunities to comment as important opportunities to contribute to the reform process.

Hence, the forestry reform process highlighted the value civil society groups put on being able to participate, even in a limited manner. Civil society recognized that the process was not perfect, but it represented a strong improvement over past efforts. Further, even though the forestry reform process did not unfold in an ideal manner, as one of the more successful environmental sector reviews during this period, it was not done quickly. The forest review process took between five and six years of formal work to produce a new administrative apparatus (the Forest Authority), a new policy, plan and legislation: a period that was much longer than anticipated, a path of policy development that was in no way linear, and a process riddled with conflict, tension and negotiation.

The forestry reform process also resembles other national policy and reform experiences in that it was clearly donor-driven. Given the historical context presented earlier, this is certainly not surprising. In the case of forestry, there is no question that the forest sector was in disarray and needing reform: forest permits and access rights were routinely used for political and economic gain. However, for some researchers, the donor-led reform agenda, even if producing technical and administrative improvements, leads to serious concerns relating to the relationship between donors, the state and citizens. Donor-led reform agendas produce 'a scenario of "donor citizens" participating in the management of a donor-driven country through processes where they use the power of their finances to create knowledge, to open and close spaces for the making and shaping of ... policy' (Ssewakiryanga 2004, p. 78). 'Taken to its extreme,' Richard Sswarkiryanga writes,

> this scenario may be seen as the emergence of a parallel state in which donors and selected central government policy actors claim their entitlements to define Uganda's route to development ... in doing so, they [donors] frequently have more influence on the way the Ugandan state functions than do its domestic citizens ... Indeed, the relationship between donors and central government actors is now a very intimate one, to the extent that sometimes a distinction between donor and government positions on a policy become indistinguishable. (2004, p. 78)

When trying to understand energy governance and the transformation of the electricity sector, these observations have important implications. If reform and policy development were the chief domain of donors, and the opportunities for consultation or participation were donor-led, who is accountable when the outcome of reform procedures are poor or when opportunities for participation are not granted? In the case of electricity, the delays and frustrations in improving the sector led to much finger-pointing, blaming and shaming, while state–society relations degraded, and access to electricity for households and firms deteriorated.

Conclusion: The legacy of reform and privatization

This chapter has painted a broad picture of reform and policymaking in Uganda from the late 1980s to mid-2000s. The central point of the chapter is to highlight how the rationale for reform in sub-Saharan Africa, along with character and culture of reform and policymaking in Uganda, related to an evolving political context. In doing so, the character of the relationship between the various actors pushing and driving reform and policymaking was emphasized. The tensions identified are not presented to reduce or ignore the enviable outcomes the country has achieved to date, such as macroeconomic stability and poverty alleviation in several parts of the country. The central point is to recognize that these achievements are a product of a specific approach and method of reform and decision-making. Drawing from several general and more specific examples – macroeconomic reform, public sector reform, privatization, forestry – some important national trends were revealed, particularly in relation to the role of various actors, the knowledge informing reforms and character of the processes followed.

The discussion in this chapter indicates that a small handful of actors were driving reform and policy choices, with donors sitting at the centre. Despite the disagreement over the degree of influence of multilateral and bilateral agencies in Uganda's early reform initiatives, it is clear the role of donors has been and continues to be substantial and dominant. Ssewakiryanga wrote about this era: 'Donors on the Ugandan policy scene are … not just funders but actors who contribute to various policy processes and are also very aware of the power that they wield in shaping policy' (Ssewakiryanga 2004, p. 83). Graham Harrison pushed this observation further: 'Rather than conceptualizing donor power as a strong external force on the state, it would be more useful to conceive of donors as *part of the state itself*. This is not just because so much of the budgeting process is contingent on the receipt of donor finance, but also because of the way programmes and even specific policies are designed and executed' (2001, p. 669). Andrew Mwenda affirmed this view: 'The World Bank is the most powerful government department' (Interview, 17 January 2003).

The extent to which the World Bank should be understood as a part of government will be examined further when discussing the electricity sector. The implication of this general observation is important when understanding 'governance' in Uganda during this period. When the role of donors is considered from the perspective of a state-oriented perspective of 'governance', the Bank's influence and authority can be understood simply as an external agent helping to guide and improve public sector management and government effectiveness vis-à-vis the creation of independent quasi-government authorities and the privatization of parastatals.

This focus on improving public management was being pushed with a clear intent to produce a better state, and what Graham Harrison suggested was the World Bank's conception of the evolution of an African state (Harrison 2005).

However, when we consider the role of the World Bank and other donors in Uganda from a relational perspective to governance, we see that donors in Uganda are instrumental to the character and quality of state–society relations. Hence, donors help promote state reform initiatives that treat participation as important, but only insofar as to produce buy-in to the reform agenda. This was achieved by extending invitations to NGOs to participate in policy development and most often in situations where the potential for controversy was minimal. As a result, despite donors assuming a role as external influences working to create more participatory processes and by default better state–society relations, donor arguments that NGOs must operate as forces of accountability, and/or contribute to policy debates, were unrealistic in the political environment in the late 1990s and early 2000s because few organizations were willing to risk challenging policy or reform proposals.

This was because: the national government was unsympathetic to challenges from NGOs, particularly those that might derail reform procedures; and, as donors usually initiated reforms and calls for participation in policy or reform, and NGOs were reliant on donors for programme funding, few NGOs could risk being on bad terms with donors if they needed financial support or wanted to participate in future policy debates. Hence, donors were intimately involved in not only pushing a reform agenda and working for greater participation in policymaking, but also in defining and shaping the character of state–society relations, and, hence, governance in Uganda, through their authoritative activities.

President Museveni also clearly played a commanding or dominant role, with donors in fact requesting him to take the lead. While Uganda's successes were undoubtedly a result of Museveni's leadership and ownership of reforms, from the perspective of policy development and policy assessment, Museveni's role as chief policy champion raised concerns. For Godber Tumushabe, the major issue was that the President was usurping the role of technical agencies in assessing policy and proposals: 'If you are an investor you go straight to State House then work down the chain of agencies; so much pressure accompanies this' (Interview, 4 March 2002). This observation suggests that the paramount role of the President in Uganda undermined the policy capacity and autonomy of civil servants and their agencies, and undermined the view that reform and policy initiatives were the result of objective analysis: policy 'decisions must be made by competent actors, not by need for political expediency ... There is a need to open up the decision-making process so that it is risky for politicians to make controversial decisions' (ibid). As I later explain, even after the major electricity reforms were complete and the Bujagali dam was under construction, conflict between independent decisions of different electricity and regulatory agencies and the President continued, particularly in relation to decisions by the Electricity Regulatory Authority to increase electricity tariffs.

As the principal public organization responsible for the monetary and financial management of the country's affairs, the Ministry of Finance, Planning and Economic Development is also central. It follows that the Minister and Permanent Secretary of the MFPED also had a central influence. Less recognized in this early period is the influence of Uganda's Parliament. Parliament was accepted as having played a prominent role in scrutinizing the pace of reforms, particularly privatization, to the frustration of many. As will be shown with the electricity sector, interviews with donor representatives and civil servants affirm the role that Parliament had in reform procedures, particularly in relation to its ability to delay passage of laws necessary for reform. Parliament, in addition to the courts, takes on more prominence in the next chapter, as their roles in scrutinizing reforms increased over time.

Given the heavy emphasis on macroeconomic reform and austerity measures in Uganda it should not come as a surprise that the knowledge driving early economic reform seems to coincide with the dominant liberal economic orientation applied and promoted in Africa. Equally, the Bank's emphasis on public sector reform as a type of 'governance reform' is also demonstrated. Despite this, specific data supporting an argument that the knowledge driving reform in Uganda was carried by a certain set of interests is not easy to discern – a link that is much more easily made for electricity. The analysis of reforms undertaken does suggest, however, that those reforms that were promoted and applied are consistent with the knowledge and ideas of international donors, particularly the World Bank. Perhaps more indicative is the fact that civil servants and elected officials that were intransigent to reforms were removed and replaced by individuals willing to lead reform post-1986.

In the late 1990s and early 2000s this chapter also shows that in practice invitations to share knowledge or to comment on government/donor proposals increased, largely due to donor requirements. These invitations, however, are not the same as processes being open to deliberate or seriously to consider knowledge or ideas generated by civil society in Uganda. While environmental NGOs and civil servants suggested that the forestry reform process represented an important improvement in the degree to which the political authorities controlling and driving reform would listen to the ideas of civil society, all actors interviewed clearly suggest that these processes are largely information sessions and opportunities to create buy-in from society.

The point here is not to demean the importance of opportunities for civil society to be heard – something that historically had not taken place. But it does raise important questions about how ideas that challenge reform proposals or question the evidence or dominant knowledge of political authorities get deliberated or considered. The evidence in this chapter is not sufficient to argue that Ugandan civil society has put forward alternative ideas for reform that were ignored by political authorities. Instead, what it demonstrates is general national trends for how reform and policymaking ideas from civil society would be and have been incorporated. The evidence revealed that civil servants and donors felt that environmental NGOs in Uganda had little capacity to produce technically proficient proposals or ideas; that civil servants and donors did not feel they have learned anything from civil society participation; and that the 'invitation culture' in Uganda limited the opportunities for civil society to put forward new ideas or to challenge government/donor proposals.

These tendencies are not unique to Uganda, and illustrate a tendency to push through reforms when they are controversial. However, these tendencies were particularly problematic in Uganda given the historically tense relationship between various constituencies in the country and because of the complexity of the reforms, programmes and policy initiatives being implemented. As a result, with respect to reform and policymaking, Ugandan civil society was caught in a paradoxical situation: they are encouraged to participate, but only if their participation does not undermine the overarching goals and/or undermine the publicly communicated linear path of improvement. As we move ahead, we will see how this pattern produced a very significant problem in the electricity reform process.

References to donor consultative dialogues in Uganda are noted in the literature; however, as the example of the process surrounding the passage of the first privatization legislation revealed, even Parliament had little influence. Hence, it is fair to surmise that the early reform period in Uganda was generally an affair of limited citizen or non-government influence or opportunity for participation. Given the extremely poor state of the country's political, social, economic and physical affairs after twenty or more years of conflict, this should not come as a big surprise. What is important to keep in mind is the political legacy of these processes with respect to the process of reform and state–society relations in reform: national reform happens more easily in the absence of debate or dialogue and when a small and select group of interests drives reform; and, hence, the process of reform is secondary to the intended outcome of reform, particularly as the potential controversy surrounding a reform is low.

To conclude, the evidence presented here regarding Uganda's reform and policymaking experience points to six key conclusions. First, controversial reform, whether successful or not, has historically been a closed process with limited opportunities for participation or debate. Second, when participation in policymaking did occur, it was usually a result of donors arguing for it or orchestrating it themselves. Third, a small ensemble of actors drove and initiated reform, with donors taking the lead, and a small collection of leaders in finance driving implementation. Fourth, environmental NGOs in Uganda had been reliant on invitations to participate in policymaking, and even when participation comes in the form of consultation or information sessions, NGOs appreciate the opportunity, but fully recognize that they have limited ability to influence reform or policy decisions as they are largely already made. Fifth, sector reforms in Uganda cannot be treated independently. The reforms that were taking place in the country were overlapping and synergistic, even if not intended to be. Forest sector reforms, for example, would influence biomass supply in the country and biomass is the dominant source of energy fuel in the country.

Finally, with respect to the complexity of reforms and the policy paths chosen, it is important to remember that World Bank country office representatives were evaluated according to the amount of programme funding they could arrange (Harrison 2001, p. 671), and therefore had an incentive to increase, not decrease the number of reforms. Further, civil servants and government communicated reform and policymaking as a linear path despite knowing that in practice this is a 'fiction' that does not take place. The outcome of this 'fiction' is that when reform or policymaking does not follow the desired path, the processes will take much longer than has been communicated and policy champions will become frustrated with delays.

When we consider that the electricity sector reforms and dam construction efforts have been characterized by government frustration with ongoing delays over review processes – most often donor-required – and NGO demands for access to information and better debate over financial aspects of projects, we begin to see how the reform and policymaking trends identified in this chapter relate to the electricity sector and how multiple transformations in the country converged to influence national ambitions for electricity transitions.

4

Dam Building
& Electricity
in Contemporary
Uganda

In 2001, the Uganda Electricity Board, the public monopoly, was unbundled, creating the Uganda Electricity Distribution Company Ltd (UEDCL), the Uganda Electricity Transmission Company Ltd (UETCL), and the Uganda Electricity Generation Company Ltd (UEGCL). As UEGCL's main functions were the operation and maintenance of the two hydroelectric generating stations in Uganda, the company was based in Jinja – the location of Uganda's first large hydroelectric dam, Owen Falls dam (now Nalubaale), and an extension to the Owen Falls dam, called the Kiira Power Station. In 2003, under the auspices of the country's privatization strategy, Eskom Uganda Ltd, a subsidiary of South Africa-based Eskom Enterprises, was awarded a twenty-year generation concession to operate the two hydroelectric facilities. Up until 2003, AES Nile Power (AESNP) – the company originally scheduled to build the Bujagali dam – also had a main office on the perimeter of the Jinja town centre. At this location, AESNP coordinated onsite activities for the proposed Bujagali dam, which was to be constructed about 10 km northwest from the town, and about 8.5 km from the Nalubaale dam complex.

I visited the AESNP offices in May 2003 to meet with the company's 'Community Interaction Officer'. The purpose of the meeting was to learn about the responsibilities and activities of the team in the lead-up to the construction of the dam – then still on track to be built in the coming year. On the morning of my meeting I made my way to Kampala's Old Taxi Park to take the Kampala-Jinja commuter bus. As the bus pulled out and made its way eastward on Jinja Road, I began speaking with the man sitting cozily beside me. Our conversation started with simple pleasantries. My companion explained that he permanently lived in Jinja but often commuted to Kampala for work. He then asked why I was in Uganda. After explaining that I was studying the reforms to the electricity sector, our conversation became livelier. My companion wanted my opinion on two issues: why does Uganda have load-shedding when it has so much potential for electricity generation; and, given the existing and available potential supply of hydro-generated electricity, why is the price of electricity so high? Why isn't UEB [Uganda Electricity Board] producing more electricity from Owen Falls? He suspected that the dam was not producing to its full potential. He went on: 'I don't understand why they chose to build at Bujagali ... tourists from Canada come to see the falls!' He accepted that the government was going to build the new dam, but he did not understand why it was taking so long to start construction.

In response to his questions, I explained that one thing that was causing the delay was the difficulty in securing financial support for the project. Foreign governments and export credit agencies were concerned about the financial viability of the project, principally whether the Transmission Company (UETCL – still owned by the Government) would be able to sell the electricity that the private company generated at Bujagali, and by default, whether there were enough consumers to purchase the electricity produced. My companion was not impressed with this explanation: 'People who want power in Uganda don't have it because of load-shedding, and Rwanda, Tanzania, the Congo and Kenya all need the power.' Thus, he was arguing that there was a large market for consumers. I explained two other issues.

First, there was controversy over the fact that people did not know how much Uganda was going pay for the electricity generated at Bujagali, and that the price of electricity to consumers would continue to increase. This comment produced a particularly strong response. Without my prompting the man argued that the high price of electricity was responsible for people degrading the environment to access lower-cost firewood and to produce charcoal, which was in high demand. He repeated that he could not understand why electricity cost so much, particularly given that 'you need water to produce electricity and Uganda has so much'.

Second, I explained that there was concern about 'country risk'; that is, uncertainty about the political stability of the country, and the impact that an unstable political environment has on investor interest. Our conversation ended rather abruptly as the share-taxi passed over the Nalubaale dam and approached the location where I had to disembark to reach the AESNP offices. I was sad to say goodbye, not least because my companion's questions and comments underscored some of the central issues that I had learned over the course of my research. His comments affirmed two simple, yet central points: (1) the public had poor knowledge of the reform process, reform choices and reform delays; and (2) there was a significant disconnect between the public's expectations about reform outcomes and donor and government goals and expectations, particularly in relation to the price of electricity and access to electricity for those not yet connected.

What was equally significant about my companion's comments was that they were repeated frequently during my early research, even when the focus of a conversation or interview was not intended to be about electricity. In more than one interview, mid-level or senior civil servants with no direct connection to the electricity sector became highly animated, visibly frustrated and emotionally charged when the issue of electricity arose. Even a Member of Parliament turned and complained to me about the electricity sector while he was haggling with the regional manager of the electricity distribution company about refinancing the debt he owed on his electricity bill! Most expressed frustration over the ever-increasing price of electricity and continuing poor quality of service provision. They stated that they did not understand what the government was doing with the sector. Most also shared a personal story about having their electricity service disconnected, or a relative having their electricity disconnected after a dispute over lack of payment or meter tampering – an event I also experienced first-hand with the accommodation I rented.

At the same time, when I told others more intimately connected with the electricity sector about the frustration people were expressing, they suggested that this was indicative of a larger problem in the country. A senior consulting engineer with Norplan – a company originally leading the development of a second dam, Karuma – and a former manager in UEB told me: '... people feel that as a middle class with good jobs and an education that they should be entitled to power [electricity]' (Interview, 27 May 2002). He further noted, however, that in Uganda electricity is not for the poor, yet 'people in power have tried to tell people that they deserve it and MPs tell them that they'll get it for them'. The question, he said, is whether 'government is prepared to talk about reality; the debate is between realism and fiction. If the issue is feasibility, then we are not talking about power to the poor. Are we looking at the reality or what's being communicated?'

In this chapter, I examine Uganda's electricity sector reforms and analyse the challenges surrounding efforts to improve the sector in the post-1986 period. As my introductory comments suggest, the post-1986 period is complex and multilayered. Further, it is not surprising that there were problems. The issue is why the problems occurred, their impact on the attempt to transform infrastructure and the economy in the country, and the political context surrounding those problems. One reason for the problems and confusion associated with Uganda's reforms relates to the evolution of domestic state–society relations. Reforms were also difficult because of *how* international actors influenced the path and process of reform, *how* the model of reform was implemented, and *how* the national government responded to domestic and international concerns surrounding the construction of a large dam – Bujagali.

In short, the chapter shows how governance challenges were central to the electricity transformation problems encountered in Uganda; as Scoones, Newell and Leach (2015a) note generally, the country is a case of what can materialize when there is confusion and disagreement over who defines the vision for change, who defines the rules for change, and what happens when different pathways of transformation conflict. Hence, this chapter reveals the evolution of energy governance in Uganda in the 1990s and early 2000s, and how the character of state–non-state interactions affected the capacity to implement reforms, provide electricity and implement a megaproject. The chapter reveals the messiness of project and programme implementation, and the relationship between this 'messiness', domestic politics, and energy and political transformations. It also reveals how the complexity of the reform exercise did not match the political and institutional capacity in Uganda. I do not argue that the reform process is the sole reason for the problems in Uganda. But I do argue that the character and process of reform and associated dam construction efforts have had significant, demonstrable impacts on domestic state–society relations, the character of governance and policymaking, and electricity provision in the country well into the year this book was completed – 2017.

The chapter begins by explaining the state of the energy sector post-1986. This section explains how Uganda's 'energy' needs were understood and what the international community suggested as necessary changes. In this section I also explain how renewable sources of energy were largely ignored in this early post-conflict period. The chapter then explains how and why public sector electricity provision under the Uganda Electricity Board (UEB) was eventually rejected, with a turn to the private sector for distribution and generation. Here I explain how the rationale and decision for electricity unbundling was reached, and how the process unfolded. Central to this explanation was the desire to construct a new large hydroelectric dam on the Nile – the Bujagali dam. As I will explain, the national government and World Bank's desire to construct Bujagali and the process surrounding the dam was a central reason for reforms unfolding in the manner they did, and by default, the problems encountered. The chapter concludes by examining the political implications of Uganda's reform experience, as well as the implications for electricity and energy provision in Uganda and future reform.

Energy in post-1986 Uganda: The fixation on dams

We can recall from Chapter 2 that between 1971 and 1986 there were no major developments in the electricity sector; Museveni's rise to power coincided with a historically low period in the generation, provision and reliability of electricity service provision in the country. In 1968, the Owen Falls dam was operating at full capacity, generating 150 MW of electricity. In 1986, its generating capacity had dropped to 60 MW. The number of electricity consumers in Uganda stood at 106,450 in 1986, but two years later the number of consumers had dropped to a low of 80,795 – a number reminiscent of the 1970s (UEB 1996). The essential problem that Museveni and the NRM were confronting was that they had inherited an infrastructure network that was very poor. But owing to the new stability in the country, demand for electricity was increasing rapidly, outpacing supply. This produced the oft-repeated cycle of power rationing/load-shedding that has plagued countries throughout sub-Saharan Africa. During this situation, the World Bank remained directly and prominently involved in Uganda's electricity sector.

In 1983, the year of its creation, the Energy Sector Management Assistance Programme (ESMAP) studied Uganda's energy sector. ESMAP advanced five electricity-specific recommendations:

1) Owing to the poor quality of infrastructure, immediately prepare and conduct a feasibility study for the repair of the Owen Falls Dam and existing transmission and distribution networks;
2) Develop a least-cost, long-term sector development program to respond to the anticipated shortfall in supply that was expected in 1988 and 1989, examine long-term demand potential in Uganda, and revisit and review hydroelectric development schemes for the Nile;
3) Owing to the role of electricity in economic growth, and in keeping with historic trends, extend Uganda's transmission and distribution network to all major towns and replace diesel generating stations;
4) Develop a second hydropower station for the purposes of exporting power, but only if commitments from purchasers can be guaranteed, and adjust current export power rates to Kenya; and,
5) Increase tariffs as soon as possible but also introduce a 'lifeline' tariff. [A lifeline tariff is also known as a 'social tariff' or 'increasing block tariff'. In this system, the first volume (block) of a service used (usually water or electricity) is provided at a lower, subsidized price, or sometimes for free. As the volume of consumption increases and passes a specific volume, the price of the good increases. Therefore, poorer consumers or households that consume a small volume of a service are supposed to benefit.]

In turn, these recommendations would provide an important foundation for all future electricity projects in Uganda, particularly those led by the World Bank. Given this list and current global attention to renewable energy it is important to note that in ESMAP's report, fuelwood, renewable energy and energy efficiency were also recognized (ESMAP 1984b, pp. 15–22). But, much to the dismay of long-time renewable energy advocates in Uganda, throughout the 1980s and 1990s the World Bank and national government's energy initiatives focused almost exclusively on electricity generation and supply initiatives. This is not surprising given the shortage of electricity in the country, but it is also noteworthy given the present attention to renewables and climate change, including in Uganda. Other bilateral agencies, such as the (then named) German development agency, GTZ, remained active in research on renewables in Uganda in the 1990s. But serious attention to the development of small-scale distributed generation sources or to fuelwood and charcoal management were overshadowed by planning for large-scale electricity generation interventions.

Josh Mabonga-Mwisaka, the Manager of the Uganda Renewable Energy Association (UREA), explained that just prior to Amin's take-over, the Forest Department was moving to make charcoal a significant economic endeavour. After Amin, however, biomass did not again emerge until the mid-1990s. But even then, he said, 'politicians didn't see the role of biomass ... it just wasn't a priority!' (Interview, 19 March 2002). As evidence of this, in 1998 Kakira Sugar Works (owned by the Madhvani Group) studied plans to install a 30 MW electric power production plant using bagasse as fuel (bagasse is the biomass that remains after sugarcane stalks are crushed). At the time, the bagasse was burnt in the open. It was estimated that the plant could have been operational in two years, providing all the electricity the sugar factories needed, and selling the excess to the UEB. The national government, however, had little interest in the initiative, and a 7 MW plant was produced powering only the factory. The explanation for government disinterest in this 'electricity from biomass' initiative largely stems from the fact that the government was focused on the development of the Bujagali project (see Nordic Consulting Group 2006). An example from the capital city, Kampala, also provides an important example of how biomass energy has been ignored.

During my fieldwork, I was introduced to an organization named Uganda Youth Voluntary Efforts in Afforestation and Environmental Protection (UYVEAEP). UYVEAEP was notable for having led the development of an innovative community programme that combined community waste collection, management and energy. UYVEAEP's operations were based in the Parish of Kasubi, Rubaga Division, in the northwest corner of the District and City of Kampala. This was one of the poorer and denser parts of the city and suffered from poor waste management and drainage problems. The members of UYVEAEP included a doctor with a clinic in the parish, several young men and women, some with university education, and several volunteers.

Originally UYVEAEP began trying to plant more trees in several zones of the parish. In a short time, however, they realized that the problem of waste management and collection was a major challenge in the area, and turned their attention to a community waste management project. Two other organizations provided UYVEAEP with some small funding to support their efforts. The Shell Uganda Foundation funded coveralls and equipment, and Living Earth Uganda, a national environmental NGO, provided funding for UYVEAEP to rent a small plot of land in the parish, which would be used to create a community waste management centre. UYVEAEP was also in regular contact with one Kampala City Council (KCC) employee who provided moral support and strategic advice.

Through this initiative, UYVEAEP rapidly expanded its activities. Soon it was providing twice-weekly door-to-door household waste collection for residents in three of the nine zones of the parish. Residents paid a small fee to have their waste collected. The money went to the twenty youth volunteers who were involved in the waste collection. Residents would also independently bring waste to the waste management centre, where the waste would be sorted into organic and inorganic material. This practice removed two of the most problematic waste streams in the city – food waste and plastics, particularly plastic bags, which blocked drainage channels and were repeatedly blamed for causing flooding within the city. UYVEAEP did two things with the organic waste: (1) it began composting in the hopes of producing fertilizer; and (2) it used matooke peels (green banana) to manually create charcoal briquettes.

The community initiative was impressive for several reasons. First, it addressed a pressing waste management problem in the city – collection problems. In 2003, the Kampala City Council did provide waste collection. But collection was only from various large garbage skips spread throughout the city, which residents had to walk to dispose their waste. Owing to the need to walk to the skips, and the poor conditions of the skips, residential waste in the city was often burned or dumped. UYVEAEP was so successful in its waste management efforts that the KCC demanded that UYVEAEP halt its activities because even while sorting the waste it collected, it was filling up the nearby skips too quickly for the KCC to empty them. Second, UYVEAEP was employing many youth, particularly young men, who otherwise had no work. Third, UYVEAEP was taking a form of biomass – green banana or matooke – widely used in southern Uganda and forming a large component of the urban waste stream, and turning it into a very good alternative, reliable energy source that could easily be used in common charcoal stoves. Moreover, the quality of the briquettes was confirmed by UYVEAEP and local residents.

Given that 95% or more of Uganda's national population was then relying on biomass for cooking, that the availability of woody tree biomass was declining, and the technology to produce briquettes from biomass waste was simple and easily replicable, this type of community endeavour offered a unique opportunity to address multiple urban concerns relating to human health, environmental quality, and social and economic well-being. Nonetheless, UYVEAEP's activities were not supported or replicated despite charcoal briquette-making being supported outside Kampala and now done in cities throughout East Africa (Njenga et al. 2014). Indeed, bilateral donors expressed clear interest in promoting the use of biomass for charcoal in Uganda in the early 2000s. At that time, however, they had only supported projects outside the capital city and only discussed rural populations that had adopted charcoal briquette-making activities due to a shortage in wood fuel.

When asked about the promotion of charcoal briquette technology the Ministry of Energy's Assistant Commissioner for New and Renewable Energy said that historically, energy in Uganda has always been about petroleum and electricity, and that biomass was not considered in relation to energy until the mid-1990s (Interview, 23 May 2002): 'It is still an uphill battle to get government to focus on it [biomass] ... there is a lot of bias into hydro.' In turn, government disinterest in biomass also translated into private sector and non-profit sector disinterest. Of the 43 private companies the Uganda Renewable Energy Association (UREA) represented in the early 2000s, most were focused on a nascent solar power market, with little interest in charcoal or biomass as an energy source. Mr Mabonga-Mwisaka, head of the UREA in the early 2000s, argued that a well-organized charcoal industry could have served as an important source of income and employment, particularly for the number of young unemployed boys in the country. While he knew the Bujagali dam would provide the government with export earnings, he wondered how it was going to help with poverty. Other non-government officials echoed these sentiments. Hence, together, the examples of the Kakira Sugar Factory and UYVEAEP and UREA's experience help illustrate the extent to which energy in Uganda was equated with large-scale electricity generation, and furthermore, begins to hint at the weight behind the development of large hydroelectric facilities like Bujagali as a solution to the electricity transition being sought in the country in the 1990s.

In recent years, the national government and many bilateral donors have returned to biomass and small-scale electricity interventions. The turn to biomass energy is long overdue, particularly given the longstanding problems with electricity supply and distribution. A GTZ Technical Advisor working in Uganda for several years said that when he arrived in Uganda in 1999, he wanted to start work on biomass right away but the focus was on the electricity sector and reforms to the sector. After the power sector reforms, he said it was understood that a large rural electrification project was next, followed by the Petroleum Bill and then biomass. This agenda, he said, was donor-driven; the whole process was donor-driven: 'Electricity first, biomass second' (Interview, 20 May 2002). Given the significant role of electricity in economic development and the poor state of Uganda's sector, few would question this prioritization from a macro-level perspective. However, when we consider this from the perspective of providing modern energy services to the poor it is obvious that most citizens would not be benefiting from these electricity reforms in the short to long term. Another reason why attention to biomass was constantly delayed was the assumption that electricity sector reforms would move quickly, thus allowing the development of a biomass strategy afterwards. Given the problems that befell Uganda's reform efforts, and given that most Ugandans still depend on biomass energy for some if not all their energy needs, in retrospect, it is now easy to question the rationale for biomass playing such an insignificant role in Uganda's contemporary approach to energy. But it is not enough to simply say that electricity sector reforms dominated government attention. A central question is what role different actors and interests played in formulating and prioritizing energy initiatives in Uganda. When this issue is considered, the World Bank's dominant role in energy and electricity interventions emerges.

Following ESMAP's 1983 study, the World Bank sponsored three multi-million dollar sector-specific reform and improvement projects: Power II was approved in 1985 for US$28.8 million; Power III was approved in 1991 for US$125 million; and Power IV was approved in 2001 for US$62 million. These projects have been complemented by other energy-related and institutional capacity-building initiatives. Examples of other projects include: power project supplements, financial guarantees and technical assistance for the Bujagali dam, a privatization and utility sector reform project, poverty-reduction strategy papers, poverty-reduction support credits, an environmental management and capacity-building project, the Energy for Rural Transformation Project, institutional capacity-building projects, a forestry rehabilitation project, and short-term thermal power generation projects. To varying degrees Power II and III each focused on pricing, sector coordination, management, planning, rehabilitation, expansion and upgrading; at the same time, the role of the state in electricity provision was also accepted in the project designs (World Bank 2001b, p. 35).

This commitment to *public* electricity provision was reaffirmed in 1993 when the Government of Uganda, under the auspices of its policy for public enterprise reform and divestiture (PERD) programme, released a list of forty public enterprises to be divested, including large parastatals relating to banking, insurance, railways and telecommunications (Tangri & Mwenda 2001, p. 118). All parastatals were classified under five categories – retain, majority share, minority share, fully divest and liquidate. The Uganda Electricity Board (UEB) was classified a Class 1 enterprise, to be retained. This classification was short-lived, however. Six years later, a new Electricity Act was passed paving the way for the unbundling of the UEB and showing a reverse in support for the state-run company and public electricity provision.

What happened during this period to invoke this change? How did the decision to turn to privatization and the construction of the Bujagali dam emerge? And what was the effect of these decisions on reform and electricity provision?

To answer these questions, two things need to be highlighted. First, we must recall the historic trends in privatization and the privatization process in Uganda discussed in Chapter 3. Here, we learned of contrasting perspectives on the success of Uganda's early privatization experience. From the perspective of civil servants, the belief was that government was doing very well and that the chief problems with privatization rested on the fact that government and elected officials were not openly communicating the rationale for privatization – chiefly, to rid government of the financial burden of poor-performing companies. Moreover, these same individuals acknowledged that while privatization was largely supported within the bureaucracy and government, the reason for privatization was donor demands. In contrast, those critical of the privatization exercise noted their frustration with the government poorly communicating its intentions and the poor opportunities offered to participate in reform decisions.

It follows that the second issue needing to be highlighted is the relationship between the above trends in the privatization of public enterprises in Uganda and the process of reforming the electricity sector, which importantly included efforts to construct the Bujagali dam. By connecting these two interrelated processes, an appreciation for how historical, political and procedural factors converged to impede a quick and 'clean' reform process emerges. More specifically, what is revealed is that while World Bank documents largely attributed the problems with the electricity sector to public management concerns and to technical and financial problems, this assessment was incomplete. The desire to construct the Bujagali dam quickly to address Uganda's electricity supply problems meant that alternative reform options were not seriously debated. Historic rationales for building Bujagali were relied on as justifications for the project, and the success of reform became dependent on the quick execution of the dam.

An about-face: From public to private electricity provision

There is little question that in the mid-1990s the Uganda Electricity Board (UEB) was performing poorly. Several factors illustrate this. First, from the perspective of economic productivity, survey results in the late 1990s and early 2000s showed that poor electricity provision was enormously problematic for private firms. For example, a 1998 private investment survey revealed that on average, Ugandan

> firms were losing an estimated 90 operating days a year from unreliable power supply. These losses translated into high costs of production and therefore reduced the competitiveness of the private firms. The same survey found that as many as 70% of the large firms, 44% of the medium-sized firms and 16% of the small firms own a power generator. (Engorait 2005, p. 3)

A 2001 edited book presented the survey results in more detail. Ugandan enterprises identified the reliability and adequacy of electricity as the leading, and only 'major' infrastructure constraint to investment compared with other infrastructure (Reinikka & Svensson 2001, pp. 220–4). With respect to the UEB's capacity to provide individual households with electricity, the results were equally poor.

The total number of new consumers between 1993 and 1999 was fewer than 50,000 (UEB 1999), and in 1994 and 1995 the number of consumers dropped below 1993 levels, demonstrating the inconsistency in electricity supply and consumer provision. Several government ministries were also notorious for not paying their electricity bills, most notably the Ministry of Defence and the National Water and Sewerage Corporation (NWSC). Furthermore, it was well known that several Members of Parliament had not paid their electricity bills for years. One former UEB employee explained that prior to 1999, there was a dangerous trend in government and the civil service: the higher you rose in public office, the more people felt they were entitled to free services. A former UEB employee who left on his own accord told me: '… staff in UEB are not the most efficient, but the external environment is worse than the internal'. Indeed, in light of several illuminating anecdotes about the challenge of getting senior elected officials to pay for electricity, as well as my own first-hand experience, the Kampala Customer Service Manager said he 'feared for private sector investors'. In addition to the problems with the external environment, the UEB also suffered from a problematic billing system – many managers approached the need for internal improvement inconsistently or were hostile to internal improvements. In the mid-1990s, the UEB's performance was so poor that it was pejoratively known as the Uganda *Enzikiza* Board – in Lugandan, *enzikiza* means darkness.

As a result of the UEB's poor performance, and just two years after it was designated a Class 1 enterprise to be retained in 1993, interviews with several senior civil servants confirmed that 'a dialogue' took place in 1995 at which time reform and restructuring were agreed. This dialogue was formalized in an internal UEB report in 1996, which recommended restructuring and divestment. This was followed by the creation of a 'Committee on Divestiture', and eventually a 1997 Strategic Plan. In 1998, the Strategic Plan was formalized, laying out a plan for the divestment and restructuring of the UEB and the eventual passage of the 1999 Electricity Act. But in keeping with historic trends in Uganda's privatization experience, during this period (1995 to 1998), reform leaders acknowledged, 'there wasn't a lot of external participation in producing the Strategic Plan' (Interview, Emmanuel Nyirinkindi, 14 May 2002). As the Strategic Plan was the document guiding privatization of electricity in Uganda, the absence of public input is notable, but not surprising. Even private sector energy proponents agreed on this point. A Managing Director of an engineering and consulting firm explained that one of the problems in Uganda had been that its privatization processes were not transparent, including the energy sector. He said he 'knows that the [Energy] Minister is uncomfortable with the privatization process because she hasn't been open about it' (Interview, 22 May 2002).

Since the mid-1990s, donors had been asking for quantitative indices of financial and service delivery improvements in the UEB, which it was unable to provide. The most glaring illustration of these problems comes from the ratio of anticipated revenue to revenue collected. In the 1990s, at times the UEB was collecting just fifty percent of the revenue it was owed. Beginning in 1996, the UEB (and later Uganda Electricity Distribution Company) implemented multiple 'operations' or 'task forces' to disconnect illegal consumers. Variously titled 'Operation Thunder', 'Omega', then 'Sigma', these programmes were launched because of high systems losses, poor revenue collection and theft. Each programme was effective at disconnecting many illegal consumers, but in the words of one senior manager, 'Operation Sigma' was 'a propaganda thing for them'. Weekly updates on Operation Sigma's progress were published in the newspaper, emphasizing this point. It was also rumoured that a UEDCL manager was awarded a bonus for each illegal cut-off Operation Sigma performed. None of these initiatives, however, was a serious attempt to improve the public utility. In the end, it was acknowledged to me that one of the main reasons for these 'operations' was to improve the UEB's records prior to its privatization – certainly an important rationale. Overall, the public view of the UEB was very poor. Indeed, in the words of the World Bank country manager, Robert Blake, the UEB was 'dysfunctional' and 'unreformable'; its operational efficiency was almost worst in the world. 'It was amazing; the UEB was not even able to satisfy 5% of the population using electricity' (Interview, 5 May 2002).

In our interview, World Bank country manager Robert Blake emphasized that the government was unable to mobilize new funds for network expansion or improvement and the UEB could not do anything about unpaid bills from other ministries. He explained that ultimately donors were unwilling to provide more funding for the sector unless dramatic change came about. This was a point I was never able to confirm. Divestiture was the antidote deemed most appropriate by the Bank, but in the early going the government still entertained the idea of maintaining partial ownership of the utility. This idea was short-lived, however, owing to problems with previous privatization efforts where the government tried to maintain partial ownership, and more tellingly, because when the government solicited interest in the distribution and generation components of electricity under partial government ownership no firms were interested (Interview, Emmanuel Nyirinkindi, 14 May 2002).

In the end, the UEB had serious performance problems, which were impeding improvements to the sector. But these problems do not alone explain the turn to privatization. For example, as a condition of future support, the Bank could have insisted on unbundling the utility, making distribution, transmission and generation operate as independent state-owned companies, while also allowing private firms to generate electricity. This was the model eventually used in Kenya. The Bank, however, was fed up with providing support for the sector without satisfactory results. Further, given the need for new sources of generation, the World Bank determined that extensive, if not radical, sector reform could be connected to the construction of a new large generation facility – the Bujagali dam. Confirming this, Emmanuel Nyirinkindi said: 'The traditional lender [World Bank] and Bujagali were the drivers of this process.' Hence, public sector reform was part of a much more complex and ambitious vision for sector change reminiscent of the colonial period and the construction of the Owen Falls dam. This time, instead of *creating* a state enterprise to build a large dam and develop a national electricity network, the state enterprise would be *dismantled* to facilitate the private construction of the nation's electricity distribution network and generation facilities.

The logic of simultaneously combining restructuring, privatization and dam construction rested on three observations. First, the UEB, owing to a combination of internal administrative problems and political interference and the inability to raise revenue to finance new projects and expansion, could not perform its electricity distribution functions. Second, the UEB was unable to reduce significantly system losses. Third, private companies would not invest in the construction and operation of a new large electricity generation project unless its profitability could be guaranteed. For electricity generation projects, this guarantee often takes the form of a 'take or pay arrangement' whereby a government must agree, usually under the auspices of a 'power purchase agreement', to pay for a set volume of electricity, at a set rate, over a specific period, whether there are enough electricity customers to consume the electricity generated or not.

Given these conditions, especially UEB's record, and donor scepticism about UEB's potential to change, then, in the words of World Bank country manager, Robert Blake: '... restructuring fell out naturally', and the need to create the domestic conditions necessary to attract independent power producers to the generation and distribution components of electricity was revealed (Interview, 5 May 2002). Thus, the country moved forward with a 'mega-reform' or 'mega-undertaking': the restructuring and unbundling of the UEB; the development and implementation of a new Electricity Act and regulatory framework; and the construction of a large hydroelectric dam at Bujagali Falls. I use the notion of a 'mega-undertaking' to highlight the ambitious, complex, overlapping and synergistic character of these initiatives, which involved multiple complementary processes.

In a country with a very small market of existing electricity consumers, very poor infrastructure quality and weak organizational and regulatory capacity, this agenda was extremely ambitious, and its success highly dependent on the careful, consecutive and successful execution of each reform action. Most critical of all for the distribution concession was the guarantee that there would be sufficient electricity supply. For if the supply of electricity cannot be guaranteed because it is uncertain or unstable, negotiations with a private firm taking over the distribution system will be extremely difficult. Acknowledging this challenge, Robert Blake noted: 'After [the need for reform is] decided, then it becomes much more complicated and timing and sequence is important – you need to figure out how the pieces fit together.' Putting it in more specific context with respect to the Bujagali dam, one northern European donor representative carefully explained that while the construction of Bujagali was not dependent on privatization, successful privatization was now dependent on Bujagali. Indeed, he said: 'Bujagali is instrumental to privatization ... without [it] the whole restructuring of the sector would collapse' (Interview, 29 May 2002).

To acknowledge that the transformation of an entire sector is dependent on the completion of a mega-project is quite an admission for several reasons. First, in the late 1990s, there was little to no evidence that the unbundling of an electricity utility would lead to its improvement in Africa. Further, there was little experience with private sector-led construction of large hydroelectric dams. In addition, common wisdom along with recent research shows that most mega-projects are over-time and over-budget and, as earlier noted, dams routinely take 4.5 times longer to complete than expected (Flyvbjerg 2014). Indeed, Uganda's reform agenda was extremely puzzling to neighbouring countries. A senior utility manager for the Kenya Power and Lighting Company said: 'I don't know how they came to the decision to split up UEB so fast; when you find out why, please tell me' (Interview, 26 April 2002). What, then, explains why the government and donors put so much faith in this complex undertaking? In the next section I examine the rationale more closely.

Faith in privatization tested: The Bujagali dam

Bujagali Falls was identified as a prime site for the construction of a hydroelectric dam in the early 1900s. In fact, in the 1920s it was deemed the best location for a dam in Uganda but given that the Owen Falls site was easier to access, Bujagali was downgraded to a second or third best choice. A consultant's report suggested Bujagali again in 1957, but the dam was never initiated owing to the ongoing concerns (discussed in Chapter 2) with there not being enough consumers for the electricity. The site's priority re-emerged in a new study in 1986, when President Museveni came to power. It is noteworthy that several of the same consulting firms who produced the 1957 study recommending Bujagali also prepared the 1986 study. Four years later, Acres International Lt. suggested that Uganda first expand Owen Falls and then build at Bujagali. A follow-up study by Acres in 1991, titled the 'Bujagali Hydro-Electric Project Pre-Investment Study', reinforced the prominence of the project.[1]

Hence, in the context of contemporary debates surrounding the appropriateness of Bujagali as a site for a hydroelectric dam in Uganda, with the development of the site having been discussed for almost 100 years, it is important to recognize the historical weight and legacy of this locale in the institutional memory of the government and its plans for electricity expansion. Given this, any effort to challenge the merits of the Bujagali site would have to have been exceptionally strong and go beyond just pointing out potential problems. Counter-arguments or indeed counter-narratives would have to demonstrate that the cost of developing a new site would be less than the cost of abandoning Bujagali, considering the time and money already invested.

When put this way, opponents to the Bujagali project faced an enormous challenge. 'Each step along a particular path produces consequences which make that path more attractive for the next round. As such effects begin to accumulate, they generate a powerful virtuous (or vicious) cycle of self-reinforcing activity' (Pierson 2000, p. 253). Hence, '… the probability of further steps along the same path increases with each move down that path. This is because the relative benefits of the current activity compared with other possible options increase over time. To put it a different way, the costs of exit – of switching to some previously plausible alternative – rise' (Pierson 2000, p. 252).

Bujagali's attractiveness as a project rested on several factors. First, owing to the history of the project noted above, Bujagali was well known and embedded in the institutional memory and priorities of the country, as well as the World Bank. Second, in comparison to other large dam projects, the number of households and individuals requiring resettlement was low, as the immediate area affected by the dam was not densely populated. The Bujagali dam would physically displace 101 households (714 individuals). The total number of households affected (small to significant loss of land) by the dam facility was 1,288 (8,700 people). In addition, 326 households (1,522 individuals) would be displaced by the transmission lines for the dam. Moreover, even though AES and the Government of Uganda began relocating households prior to the project's approval, there was very little open resistance to, or mass protest against

---

[1] In 2002 Acres International, a Canadian company, was charged with bribery in relation to its involvement in the Lesotho Highlands Water Project (LHWP). Acres was found to have made over $2 million in payments to project officials, and was subsequently sanctioned by the World Bank in 2004, halting Acres' ability to bid on World Bank-related contracts for three years. Another engineering firm was also later found guilty of bribery in relation to the LHWP. The World Bank's Sanctions Committee found Lahmeyer International, a German firm, guilty of bribing the LHWP's Chief Executive – the individual responsible for awarding contracts.

the project. Third, the Bujagali site was also considered a good location owing to the topography of the region. The banks of the Nile were steep, making construction easier, and an island at the proposed dam location would allow the river's flow to be more easily redirected during construction. In the words of one engineer, 'if you want to build a hydroelectric dam, this is an ideal location to do it' (Interview, 22 April 2008). Fourth, according to the World Bank's 2001 Project Information Document (PID), the Bujagali dam had also been the subject of many analyses relating to project cost, environmental impacts, cultural, and socio-economic impacts, which supported its development.

With respect to the project's initial cost estimates, in comparison to other hydroelectric options, at $500 million, the project was deemed to be the 'least cost hydropower project' (World Bank 2001a). According to the Bank's report, large-scale hydropower was also 'the most viable alternative for electricity generation'. Government and Bank studies also suggested the dam would not impact the area's natural habitat negatively. It was recognized that there would be some disturbances and changes to fish ecology, but on a macro scale studies suggested that none of the nine downstream countries would observe any changes in the 'discharge pattern' of the Nile. Concerns around water levels and variation in water supply due to climatic change were not seriously debated, however, nor were climate-related concerns about the release of methane gas from the breakdown of submerged natural vegetation. Other concerns related to the cultural importance of the falls to the Basoga people.

The Bujagali Falls held cultural significance to the 2.5 million Basoga, who believed that their spirits resided in the churning waters at the Bujagali Falls (Inspection Panel 2002). Since the dam would inundate the falls, debate and discussion surrounding 'moving' the spirits took place for some time, and dispute over whether the chief priest and spirit medium, Jaja Bujagali, had agreed to 'relocation'. With respect to socio-economic impacts, one central issue that gained a lot of early attention was the loss of tourism revenue from visitors to the site and to two whitewater rafting companies that ran trips over the Bujagali Falls. In fact, some of the earliest opposition to the Bujagali dam came from whitewater rafting companies, with the general manager of one company, Adrift, being arrested in 1998 and accused of inciting opposition to the project. This short overview of general concerns is extremely simplified and does not do justice to the number of processes and volumes of reports developed to study the site's potential. But at the same time, the most critical issue here is not the details of these studies; it is how they were used to move Bujagali ahead.

The first formal steps to initiate construction of Bujagali began in 1994, when the South Africa-based Madhvani Group of companies approached US-based AES International about building the dam. That same year, President Museveni signed a Memorandum of Understanding with AES and Madhvani giving them first right of refusal to build. Together, AES and Madhvani established the company AES Nile Power (AESNP). Neither company had ever constructed a dam before. AES was not the only independent power producer (IPP) in Uganda considering electricity generation sites at the time. A Norwegian company, Norpak Ltd, a subsidiary of Norwegian-based utility company, Adger Energi AS, was also given the right to develop a dam at another favoured site in northern Uganda, Karuma Falls. (For a short time in the mid-1990s, Enron was also in Uganda.) In each case, no competitive bidding process was undertaken for dam construction at either site. Museveni's word and strength of character solidified the agreements. According to one private sector source, 'Museveni took a brave stance'; he took the reports on the hydroelectric capacity of the Nile and 'hawked them around the world', ultimately resulting in AES and Norpak's commitments. In contrast, a member of one of the domestic NGOs most critical of Bujagali, the National Association of Professional Environmentalists (NAPE), described the absence of competitive bidding as a 'silent "Scramble for the Nile"'. From either perspective, by the mid-1990s, private sector interest in Uganda's hydro-generation sites was high, with Bujagali slated to be the first site for construction, followed by another dam at Karuma Falls. Other sites were also listed for future development, including Isimba, Ayago and Kalagala – all part of the Nile system and several being developed.

The Government of Uganda's initial desire was to construct the Bujagali and Karuma dams at the same time. Together, the two projects would add approximately 350 MW of electricity to Uganda's grid, thus doubling the volume of electricity available in the late 1990s. Given the inability to store electricity and the absence of a large enough domestic network and established export market for this volume, the Bank felt that together both dams would add too much electricity at once. This argument echoed the World Bank's rationale not to build a dam in Kenya in the late 1950s – it was believed the Owen Falls dam would provide sufficient supply for the region. Despite this, by 2013, some were estimating that Uganda's demand would surpass supply by 2015 (Rotberg 2013, pp. 119–21). In a presentation at the World Bank's 2006 Energy Week, Uganda's Energy Minister until 2006, Syda Bbumba, paraphrased the Bank's position this way: 'Uganda's macro-economic stability would be overturned by such massive investments. So, our development partners *forced us* to settle for one project [Bujagali] ... Did Uganda have the demand to warrant even one project of 250 MW? On this, unending studies were carried out by the World Bank Group' (Bbumba, presentation, World Bank Energy Week, 6 March 2006, emphasis added).

The other rationale given for beginning with Bujagali was that the project was further along than Karuma. But on this point, consultants working on the development of the Karuma project disagree, and environmental groups questioned why other generation options such as geothermal energy were not better studied. Indeed, as far back as 1982 there were studies suggesting Uganda could develop nearly 450 MW of electricity from geothermal sources (McNitt 1982 in Bahati & Natukunda 2009). While studies of geothermal electricity potential existed in the 1990s and early 2000s, none of this potential was developed, and in 2016 no geothermal projects were operational in Uganda, although a great deal of exploration continues. This contrasts with Kenya, for example, where 51% of Kenya's current electricity supply now comes from geothermal sources (World Bank 2015a). Kenya continues to develop new geothermal sources in the hopes of minimizing the reliance on electricity from dams owing to unreliable rainfall and water to maintain hydroelectric reservoirs.

With respect to whether to develop Bujagali or Karuma first, Mr Lawrence Omulen, the Managing Director of Norplan Ltd, the firm that was working on behalf of Norpak Ltd to prepare for the construction of Karuma in the early 2000s, noted that AES was in Uganda in 1995, but Norplan was in Uganda by 1996 working on the Karuma project. Karuma Falls, located in north-central Uganda, would be a much smaller dam than Bujagali, at 100 MW as opposed to Bujagali's 250 MW. Mr Omulen argued that there were several reasons why Karuma was a better project than Bujagali: the number of people that would lose their land and have to be displaced was lower for Karuma (200 people) versus Bujagali (700); the hydrological risk – the risk that the supply of water needed to generate the expected electricity output would be unavailable – was virtually zero; the dam's proximity to northern Uganda was very advantageous owing to poor access and poverty in the north, and the decreased transmission losses resulting from the electricity having to travel a shorter distance; and, in contrast to criticisms of AES (noted below), Norplan had always been open about the cost of its project and the price of electricity it was expecting from the sale of electricity generated.

It is noteworthy that the rationale for putting a dam in a region with high levels of poverty and low industrial activity, northern Uganda, assumes a poverty-oriented, distributional justice rationale for site selection and dam construction. This perspective did not take hold, however, and Bujagali remained a priority. Norplan also argued that it had completed all of the required studies it needed to before Bujagali, and despite the usual controversy around such practices, had also compensated and resettled all of the population surrounding the Karuma site prior to AES doing so for the Bujagali site.

While Mr Omulen's perspective may be considered biased given his role in the development of the Karuma project at the time, it is noteworthy that opponents to the Bujagali dam were *not* against the Karuma project – there was no blind opposition to dams. For example, when I asked Martin Musumba, member of the small organization, Save Bujagali Crusade, if he would support Karuma as the first dam project in Uganda, he replied: 'I would endorse that' (Martin Musumba, Save Bujagali Crusade, 9 May 2002). This perspective is important because the national government's frustration with 'environmental groups' was mostly related to groups who opposed Bujagali. Indeed, Syda Bbumba wrote in her World Bank Energy Week presentation: 'Bujagali became a subject of demonization by environmental groups which had another agenda. This debate was allowed to derail the implementation of the project for seven years to the point where the developer, AES, which was also experiencing a financial squeeze at the time, decided to withdraw.' Before considering the project in more detail, it is important to acknowledge how Bujagali and the electricity reform process were linked and the interplay between the World Bank, other donor agencies, and domestic and international NGOs.

The politics of process: Understanding and explaining delays
The debate over constructing Bujagali or Karuma did not last long. Once Bujagali had been prioritized, the Government focused its efforts on implementing the institutional and legal conditions necessary for private firms to generate electricity in the country. From the time of the creation of the Uganda Electricity Board (UEB), it was the only company legally permitted to generate, transmit and distribute electricity in the country. And while parliamentary debate suggests there was some controversy over whether the Electricity Act had to be amended to permit new companies to generate electricity, an amendment to the Act was pursued, and by 1999 a new Electricity Act had been promulgated. With the new Act coming into force in 2001, the UEB was split into three separate companies, with distribution and generation being prepared for privatization.

Internal resistance to the unbundling of UEB did exist, but it was muted by the momentum of the entire reform and dam construction endeavour. Hence, while the regulatory and service delivery structure for electricity was being changed and implemented so too were plans moving forward to build the two hydroelectric dams. In fact, while AES and Norpak were developing their hydroelectric projects they were also providing advice to government agencies on how to change the institutional and legal framework for the electricity sector. For example, during the late 1990s the National Environmental Management Authority (NEMA) was developing new legislation for environmental assessment procedures while AES was planning Bujagali – a project that would eventually be scrutinized using the EA legislation. Senior members of Norpak also told me that they held workshops with MPs to explain what a power purchase agreement was – the proprietary agreement between a private firm and the government identifying the terms of the management and sale of electricity. Moreover, while AES was pursuing one of the largest private sector investments in Africa at the time (approximately $550 million), and Norpak was waiting to start construction on Karuma, three principal government agencies relating to energy issues – Forestry, Energy and Environment – were all being reformed.

What makes these events even more interesting is how the World Bank Group became involved in Bujagali. In 1991, the Bank's third power project in Uganda, 'Power III', was approved. Owing to poor electricity supply and poor infrastructure quality, one of the central components of this project was the addition of an extension to the Owen Falls dam in order to add upwards of 200 MW. The extension, Kiira, was not complete until 2000 – a period of time much longer than anticipated. Despite its completion in 2000, additional generation units were being added to Kiira well past this date owing to the fact that the extension only added 100 MW to the grid initially.

The delay in initiating the extension project spurred the Government of Uganda to look for other generation options at the same time that Kiira was under way. Hence during work on the extension, in 1994 the government turned to Bujagali, and in the same year guaranteed the site to AES. Shortly after this, AES asked the World Bank to provide direct financing for the project and to help with additional financing. According to individuals within NEMA, however, the Infrastructure Finance Corporation (IFC) began reviewing the project without their knowledge. Around this same time (1995–96) discussions had also begun relating to the unbundling of the UEB. Subsequently, in 1997, the Government of Uganda requested a Partial Risk Guarantee from the IDA to support the development of Bujagali.

Bearing in mind the earlier discussion of the timeline associated with the unbundling of the UEB, there is clearly strong evidence to suggest that the World Bank was formally involved or at minimum had direct knowledge of the development of the Bujagali site prior to or while it began promoting the unbundling of the UEB. Events and documents show that the Government and most directly the President were taking a prominent lead in the development of Bujagali. An unprompted remark by an AES representative in a 2002 interview confirmed the general feeling that Bujagali was as much a World Bank project as it was a Government of Uganda project: '... the World Bank is really the proponent of the project', and was taking the lead in negotiations with export credit agencies to coordinate the project's financing (Interview, AES staff member, 21 March 2002). Thus, the Bank played a central role in blending public sector reform goals with private sector participation and dam construction. The extent of this role is revealed further when we look at NGO concerns with the project, along with the problems encountered.

In 1998 domestic civil society organizations began to question publicly the rationale behind the construction of the Bujagali dam. In the early going, two organizations took the most interest in the project: the National Association of Professional Environmentalists (NAPE) and the Uganda Wildlife Society (UWS) – both domestic environmental NGOs.

NAPE was established in 1997. In 2002/03, it had six permanent staff and 65 registered members. Members required a diploma or a degree in environmental science. An Executive Director, Frank Muramuzi, led the organization. The UWS, established in 1993, was also a membership organization, then with nine permanent staff. Both organizations were based in Kampala and had been involved actively in environmental policy development and advocacy.

In 1998, NAPE visited the Bujagali site and spoke with some of the residents who were going to be resettled. NAPE Executive Director Frank Muramuzi, and NAPE representative Geoffrey Kamese suggested that despite the general perception that the local community was in support of the project, 'the people that challenged the dam were suppressed' (Interview, 11 March 2002). The extent of discontent, however, is hard to determine given the fact that by this point AESNP was already engaged in environmental and community assessments, was speaking to affected residents, and was thus bringing out expectations of resettlement and providing compensation. Hence, I could not confirm or refute the community support or resistance to the project by the time I started studying the project. This same year, 1998, the UWS organized a workshop to discuss the Bujagali project as there was mounting concern that Parliament was being pressured to approve the Power Purchase Agreement (PPA) before the environmental assessment of the project was complete and approved. One year later, in 1999, the Berkeley, California-based environmental advocacy organization, the International Rivers Network (IRN), now named International Rivers, became engaged in the Bujagali project.

Created in 1985, International Rivers (IR) is one of the most important transnational anti-dam advocates. IR facilitated communication between dam-affected communities and international actors, and had 'sufficient technical and analytical capacity to credibly challenge and interpret claims about the costs, benefits, and effects of large dams' (Conca 2006, pp. 176–7). As a result, connecting with IR was a very important strategic decision and opportunity for Ugandan environmental NGOs, whose advocacy and national policy influence was weak.

NAPE and one other small, unregistered Ugandan group, Save Bujagali Crusade (SBC) – an umbrella coordinating group – approached IR for assistance owing to the NGO's prominent global role in advocating against large hydroelectric dams and for the protection of river systems. Neither advocacy group received direct budgetary support from IR, but they were now connected to an international organization that had an important international profile, that had intimate knowledge of the World Bank and the private dam-building industry, and access to information that they could not easily access on their own.

One of the most important sources of information was the Washington-based Bank Information Centre (BIC). The BIC provided access to World Bank documents and ensured that the NAPE and SBC were present at World Bank stakeholder meetings held in Washington, DC and international meetings of the World Commission on Dams (Linaweaver 2003, p. 291). The connection with IR also raised the profile of NAPE in Uganda (ibid), which was the lead anti-Bujagali group in Uganda and one of the few that made Bujagali their primary focus.

Other NGOs like the Uganda Wildlife Society, Advocates Coalition on Environment and Development (ACODE), Joint Energy and Environment Program (JEEP), and Living Earth Uganda, were also engaged with energy issues but were not focused directly on large-scale electricity infrastructure as a dominant programme issue. Most focused on ecological and environmental issues.

Moreover, it is important to note that NAPE and SBC's attention was focused on the dam, and not electricity as a service or privatization as a practice. This is important to keep in mind when reflecting on the strategies of Ugandan advocacy groups. For if NAPE, SBC and others had framed their concerns around electricity, the experience of other advocacy groups in sub-Saharan Africa, such as the Movement for the Survival of the Ogoni People (MOSOP) in Nigeria, suggests that they would not have been as successful in gaining international attention (see Bob 2001). This is because of how important it is that a 'fit' exists between domestic NGO concerns and the purpose and focus of international NGOs. If a fit does not exist, then international NGOs will be reluctant to invest time that is unlikely to help advance their own international campaigns or is outside of their scope or expertise. Few if any international NGOs were making a prominent name for themselves arguing for greater access to electricity in the 1990s and early 2000s, perhaps apart from unions and community organizations in South Africa and Nigeria.

Originally, the central concerns NAPE and SBC advanced about Bujagali related to the dam's social impacts and the ecological impact on the river system. Climate change and water levels were also raised as concerns, but at the time Bujagali was being initiated in the early 2000s, historical river flow records suggested the run-of-the-river dam would not impede the volume of water available or flowing downstream. By 2006, however, at the height of Uganda's electricity crisis, Uganda was accused of furthering water level drops in the Lake Victoria Basin. The government was accused of ordering too much water be released from the two existing dams in order to curb the dismal availability of electricity. The release of water to generate more electricity combined with a drought in 2006 converged to cause grave concerns for water levels in the basin and Nile.

Following this period, and since then, with recurring droughts in East Africa, concerns about climate change and a reliance on large hydroelectricity have grown significantly (see Alstone, Gershenson & Kammen 2015; Kammen & Pottinger 2015). Kammen et al. (2015), for example, argue that climate change is expected to dramatically affect the power sector and that the loss of hydropower capacity on some river systems could drop by one half. As is explained below, despite these concerns, Uganda continues to pursue the construction of several large hydroelectric dams while also increasing the number of small-scale distributed, climate-resilient renewable generation sources with a large amount of support from European donors and the World Bank.

Turning back to Bujagali and the advocacy strategies used by domestic and international NGOs in the first half of the 2000s, in a short period, and with the advice and support of IR, the central concerns became more focused. Four issues came to dominate the debates over Bujagali: the project's cost; the impact of the project's cost on electricity prices; access to and disclosure of information about the project; and whether alternatives to Bujagali were adequately considered. The refinement of these concerns and the increased sophistication of NGO advocacy and analysis were observable over the two years that I interacted directly with NAPE during my initial fieldwork and in the years that followed.[2]

One of the first concerns that Ugandan NGOs highlighted related to the intent of the Bujagali project and its beneficiaries. They argued that the primary purpose and benefits of the project were never accurately communicated. While government officials publicly stated that the project was a poverty-reducing measure, the reality, as interview data with NGOs, donors and government officials confirmed, was not to supply electricity to individuals but to fuel industrialization. Thus, the 'poverty impact' of the project would be indirect; the goal was to fill the supply gap in order to fuel industrial and economic development and not to immediately connect households. This model, in and of itself, was not really the central problem. The issue was that Ugandan citizens were led to believe that the dam was going to benefit them directly: a point confirmed by my informal conversations with lay citizens. Moreover, as Stephen Linaweaver (2003, pp. 288–90) highlights, the Busoga kingdom, the kingdom on the east of the Nile near Jinja and beside the Bujagali construction site, felt that the local district government had not been properly consulted and that AES representatives had made false

---

[2] Indeed, demonstrating their lead role in debates about dams in Uganda, in October 2004, NAPE was chosen as the Secretariat for the Uganda Dams Dialogue, with the Ministry of Water, Lands and Environment in the chair. The dialogue process was funded by UNEP. The Dialogue is a model that has been used in other countries in Africa, such as Nigeria, South Africa, Togo, and Ghana, and builds on the work and recommendations of the World Commission on Dams (WCD). Details of the dialogues can be found here as of May 2017: http://staging.unep.org/dams/. Country studies aimed at generating consensus on dams and development issues and to make recommendations to strengthen decision-making regarding dams and development like the WCD process. Uganda's Steering Committee comprised representation from local governments, national government ministries, private sector, and civil society. The World Bank, UNEP, and GTZ had observer status.

promises to give them free electricity, free hospitals and free education. The Basoga's concerns were informed by their historical experience with dam construction in Uganda. Following the construction of Owen Falls in 1954, the Basoga were also promised electricity, which they never received.

The second central concern of NGOs related to the cost of the project and the cost of future electricity. As I highlighted at the start of this chapter, for many citizens the assumption was that the cost of electricity would *decrease* with the completion of the dam. The reality, however, again not communicated, was that the total price of electricity for consumers would *increase* dramatically initially in order to pay for the project, for transmission and distribution infrastructure and delivery, and for the favourable return AESNP was to receive on its investment. Even those supportive of the dam and intimately aware of Uganda's electricity sector acknowledged what was a stake: 'AES is not bringing anything to Uganda and that all this talk about a $550 million invest-ment is rubbish – AES will make a lot of money in Uganda' (Interview, Paul Maré, former Managing Director Uganda Electricity Board, then Managing Director Eskom Enterprises Uganda, 17 January 2003). At the centre of concerns surrounding the project cost, and one of the most contentious issues surrounding the first effort to construct the Bujagali dam (Bujagali I), were the financial details of the Power Purchase Agreement (PPA) between AESNP and the Uganda Electricity Board, as well as the conditions imposed on the GOU that were imbedded in the Implementation Agreement (IA) with AESNP.

A Power Purchase Agreement is a long-term contract for the sale of electricity to one buyer. In the case of Uganda, the seller was AESNP and the buyer was originally UEB, then UETCL. PPAs usually run for 20 to 30 years, during which time the private investor expects to receive an agreed and regular return on their investment. PPAs also stipulate other necessary agreements, such as dispute resolution mechanisms and agreements relating to failure of payment. The intent of a PPA is to mini-mize the private sector's risk in investment. In the case of hydroelectric dams, this risk could include decreased water supply or construction problems. In addition, PPAs ensure stability in the investment return over the course of the agreement (in Uganda the PPA was for 30 years) whereby the annual payments and tariffs remain relatively stable for the course of the agreement. This is to protect against such things as economic change, decreased demand for electricity, or political change or insecurity. As Peter Bosshard (2002) explains,

> Private investors are often not prepared to accept these substantial risks … projects will only go ahead if governments are prepared to assume them. So-called take-or-pay clauses require the 'off-taker' (the utility buying the power) to pay for a pre-determined amount of electricity from a hydropower project even if the plant is unable to

generate this amount because water flows are inadequate, and even if there is insufficient demand for the power from consumers... Power Purchase Agreements also define the prices a utility has to pay for the power produced by the project, or rather for the project's capacity to generate power. The prices are supposed to reflect the distribution of risks between the government and the sponsor. They need to cover the debt service for project construction, the operation and mainte- nance costs, taxes, and the return on the investor's equity. (Bosshard 2002, p. 6)

At the centre of debate in Uganda was whether the PPA should be made publicly available.

NGOs argued that because the PPA was negotiated with the Govern- ment of Uganda, it was a public document that should be made available. Government officials countered by explaining that the PPA contained proprietary information and could not be disclosed. Later, government officials informed me that the PPA had been made available to Members of Parliament in the 6th Parliament and was available in the Parliamen- tary Library. World Bank country manager Robert Blake concurred that it had been made available to the 6th Parliament. I went to the Library on several occasions never to find the document available. A northern European diplomat explained that MPs did have access to the PPA, but that it was a condition of Parliament that it not be disclosed to anyone on the 'outside'. Eventually, however, this debate turned moot. A Ugandan NGO, Greenwatch, went to the High Court and challenged AESNP and the government's argument that the PPA was proprietary. The Ugandan High Court ruled in favour of making the agreement public. This is one important indicator of the political and institutional transformation that was emerging in Uganda in the mid-2000s, which the government and perhaps the World Bank were not anticipating: public institu- tions were willing to challenge government preferences when they ran counter to established legal principles. Despite the important role of the High Court, even prior to its decision, a copy of the PPA was leaked to several civil society groups.

The outcome of the document's availability meant that the financial conditions of the agreement were now available for analysis. With this information in hand, NAPE shared the PPA with International Rivers (IR), and IR sent the PPA and the Implementation Agreement (IA) to India-based Prayas Energy Group for analysis and assessment. From the 35-page assessment and review by Prayas, two broad conclusions about the PPA are worth citing:

The World Bank analysis of the PPA ... is substantially weak. At times it contradicts the actual provisions of the PPA. It also fails to highlight key issues such as the high capital cost of the project, the risks of possible high debt cost, the risk of very low liquidated damages to UEB in case of construction delays... The PPA is also substantially unfavorable to UEB and the Ugandan government on several other accounts. For example, the PPA requires the government to restructure UEB, limits the control of UEB and the government on the financing and other contracts of the project, and grants AESNP a right of first refusal even after UEB has repaid all the equity, including returns, of the project. (Prayas Energy Group 2002, p. 29)

Added to this, Prayas further revealed the extent to which the Bujagali project was driving UEB's reform and unbundling. Under the heading, 'Conditions imposed on the government', Prayas noted that the IA in the PPA required the government to accept certain obligations. 'First, the IA require[d] GoU to prepare and complete an implementation plan for either the privatisation or capitalisation of UEB, and the commencement of such an implementation plan. This first provision shows a clear belief that privatisation is essential for an improvement in performance of the sector' (2002, pp. 12–13). The second important obligation related to the government's capacity to enter into a new Power Purchase Agreement with other firms: 'According to the IA, GoU / UEB are prevented from entering into any new PPAs or IAs for other projects until AESNP attains Financial Closure, unless they can expressly and independently provide evidence that such new projects are financially sustainable without affecting GoU / UEB's ability to sustain the Bujagali project' (2002, pp. 12–13).

This provision, therefore, explained why Norplan was not able to proceed with the Karuma dam until Bujagali was well advanced. According to Prayas, in theory, this provision made financial sense. But this obligation could also produce a serious problem. If there was a delay in the implementation of Bujagali owing to any number of factors, for example a contractual dispute, the government would be unable to sign a new PPA for another project. As Prayas wrote, '[T]his is *extremely risky ...* and could have serious implications for the future power supply scenario in Uganda' (2002, p. 13, emphasis added). Later in its report, Prayas wrote more critically: 'It is deplorable that IPPs [Independent Power Producers] force such policy decisions on developing country governments. It is even more deplorable that the World Bank actively supports and encourages such provisions and projects' (ibid, p. 26). Whether one shares the same level of concern for this practice as Prayas, the conditions in the IA clearly illustrate the extent to which the unbundling of the UEB was directly tied to Bujagali and was endorsed by the Bank.

As the above information would suggest, there were also serious concerns about access to information and transparency, along with the character and opportunity to participate in the dam's assessment and review process. Here an important debate arises over forums or spaces for debate and deliberation. On one hand, the Bujagali review process, particularly the Environmental Impact Assessment process, was deemed an international best practice; it was thorough, expensive and seemed to comply with World Bank and Government of Uganda requirements (see Linaweaver 2003). The issue, however, was not how many times consultation took place, but the character of participation – the quality of the space to engage in the process. One public meeting held in Jinja in 1999 demonstrates this point.

The meeting was a raucous event described by pro-dam and anti-dam groups as 'havoc', a 'terrible situation' and 'very mismanaged'. Proponents and opponents of the project charged each other with paying off participants. Moses Isooba, with the Uganda Wildlife Society (UWS) at the time of this Jinja public meeting, said that AES and the government thought that the concerns raised by NAPE, UWS and Greenwatch were in opposition to the project. 'No!' he said, 'this was about the process and procedural issues; but, asking procedural questions meant opposition to the project' in the eyes of the government and AES (Interview, 11 March 2002). Mr Isooba went on to say that the World Bank stated that they had 'never seen someone consult like AES', and that this was one reason why they approved the project. But with respect to the consultation process between AES and the dam-affected communities and the content of these consultations, Mr Isooba said that questions about procedure, alternatives, access to information and project costs 'aren't questions that my mom and dad will ask from the village'. Mr Isooba stated categorically that Ugandan NGOs were not anti-dam; they were demanding that 'the procedure be properly scrutinised'. In addition, with respect to the factors addressed in the EIA, it is important to note that this process was restricted to social and ecological issues. Scoping an EIA in this way may be sufficient if there is an alternative means for the public to participate in the economic review of the project, but the EIA did not address project costs, and there was no similarly 'thorough' 'best practice' for the economic analysis.

A final central concern was whether a thorough analysis of alternatives to Bujagali had been done. On this point, NGOs pointed to the Karuma dam and geothermal power sources. With respect to the Karuma project, the ecological and social impacts from this dam project were much less than Bujagali. Karuma is also located in northern Uganda. Therefore, in comparison to Bujagali and with respect to electricity transmission losses and increasing access to electricity, if the goal was to provide grid-based electricity to households and businesses in northern Uganda, Karuma's proximity to northern settlements would help reduce transmission system losses and increase the potential to increase northern access to electricity; however, as earlier noted, the goal of Bujagali was to meet growing demand, which was primarily in southern Uganda. The unit price of the electricity generated from the Karuma was also going to be less than Bujagali. However, it was acknowledged in interviews with NORAD representatives and also by Stephen Linaweaver's research that one of the reasons why Bujagali trumped Karuma was the intense lobbying effort put forward by the US

Ambassador at the time. Linaweaver reports that in his effort to quiet dissenting MPs, President Museveni told Parliament: 'The US government had warned that a failure to approve the dam would threaten Uganda's relationship with the world's only superpower' (2003, p. 293). Moreover, as noted above, the Bujagali Implementation Agreement (IA) required financial closure on Bujagali before the government could begin negotiations with other IPPs. Therefore, even if the Government of Uganda wanted to start Karuma, it was unable to under the conditions written into the power purchase agreement with AESNP.

But this debate also highlights the mixed narratives surrounding the rationale for the project and the intended purpose of electricity generally. If the government had articulated that the central purpose of electricity reform and Bujagali was to first meet industrial and commercial demand, then the dam at Bujagali made sense owing to its southern location where population and businesses were concentrated. Instead, there was a general narrative circulating among the public that households would gain access to low-cost electricity after these dams were completed. While this was not true, it was understandable that the Government of Uganda did not state this publicly. In addition to Karuma, Ugandan NGOs also argued that the geothermal potential of the country should be pursued.

On the subject of geothermal, World Bank country programme manager, Robert Blake, said that the Bank did look at geothermal power, but that the cost of developing it was 'very high' and 'no one had proved otherwise'. He said that the Bank was not convinced they could produce geothermal efficiently, while the choice of Bujagali was 'a no brainer. If the costs of geothermal are less than Bujagali then forget about it; go with geothermal.' But reinforcing earlier comments about increasing returns, Blake explained that the cost of preparing a geothermal project had to be weighed against the money already invested in the Bujagali dam. He said that the project proposal for Bujagali was already complete after having been studied for some time, so the cost of geothermal had to take into account the cost of preparing a new project proposal, in addition to the project itself.

Owing to the number of concerns raised by IR and NAPE, in 1999 Parliament and the World Bank delayed a decision on the dam. The height of conflict, however, emerged in 2001. AESNP began compensating dam-affected communities prior to World Bank financial approval, and prior to AESNP having the necessary $115 million equity it needed to move ahead with the project. Owing to these and the above-noted concerns, and after writing to World Bank management and receiving an unsatisfactory response, in July 2001, seven individuals – one from NAPE, two from Save Bujagali Crusade – filed a complaint with the International Finance Corporation's (IFC) Compliance Officer and the World Bank Inspection Panel. (The Inspection Panel is the independent body established in 1993 by the World Bank Executive Board owing to the protracted controversy over its financing of the Sardar Sarovar Dam in India.) The request for inspection focused on several Bank operational policies the NGOs felt had been contravened, particularly in relation to environmental and financial factors. The request for inspection was made four months prior to the date when the Bank Board was expected to vote on the project for approval. The Bank management responded to the issues raised in the Inspection Panel request, but the NGOs were again not satisfied with the response. Subsequently, in early October 2001, the Inspection Panel recommended a full investigation of the allegations, and at the end of the month the Bank's Board of Executive Directors approved the recommendation authorizing the Inspection Panel's investigation. The Inspection Panel began its investigation shortly thereafter.

Frustrated with ongoing delays, while the Inspection Panel began its work, President Museveni was also taking action to try to ensure that the Bujagali Project would continue. In October 2001, he wrote to World Bank President, James D. Wolfensohn, to request the Board make a decision prior to the conclusion of the Inspection Panel's review (Linaweaver 2003, p. 292). This was a culmination of President Museveni's influence on the Bujagali project starting with the allocation of the Bujagali site to AES, the clear understanding in government ministries that that support for the Bujagali project had come down from the executive, and that without the President's intervention the project would not have come as far as it had (ibid, p. 293). It followed that on 28 December 2001, the Bank Board approved the Partial Risk Guarantee for the project. The rationale for not halting a decision rested on the fear that Ugandan groups would use or abuse the Inspection Panel process to delay projects. Shortly after, on 24 January 2002, less than a month after World Bank Board approval, the groundbreaking ceremony was held with the clear assumption that the project was now moving ahead. But, one year later, in 2003, estimating a loss of US$75 million, and amidst allegations of corruption involving project sub-contracts, AES withdrew from Uganda, and the Bujagali I project temporarily halted.

Problems in implementation: Ambition, conditions and complexity
In October 2005, the World Bank wrote a Project Completion Note assessing Bujagali I. The note argued that the project failed for three reasons: (1) a withdrawal of export credit agency support due to the high level of perceived country and business risk in January 2002; (2) ongoing investigations and allegations of low-level corruption involving one of the construction contractors; and (3) the deterioration of the private sponsor's (AESNP) global financial situation, following Enron's collapse and a loss of confidence in high-risk global energy undertakings (World Bank 2005b). The analysis emphasized financial and technical problems. At the end of the Note, under the heading 'Lessons Learned', the

Bank wrote that for any new generation project, the Bank would take note of the Inspection Panel report and Management's response to the report, along with the importance of: (1) a robust financing plan; (2) 'a transparent and competitive process for the selection of the civil works and electro-magnetic equipment contractors' (a reference to the corruption charges against a dam sub-contractor); and (3) 'ensuring the efficient operations of the power sector's distribution business including improved quality of supply and access' (ibid, p. 4). Absent from these explanations or lessons (but perhaps implied by the statement that 'the Bank take note of the various issues raised by the Inspection Panel') was any specific reference to the host of procedural problems encountered in the dam-building exercise, the relationship between the dam and the complexity associated with sector reforms, and the demands the reform placed on the government and civil service.

Financial problems were clearly instrumental in the dam's initial delay. One cannot say that if the global financial outlook was better at the time that AES might not have pulled out and the dam completed. But pointing to the problems with Bujagali as largely technical and financial, as the World Bank did, or critiquing the Ugandan Parliament or domestic or international NGOs for undermining Bujagali (see Mallaby 2004) pays insufficient attention to the process of dam construction and the sector reforms in which it was embedded. Indeed, in the Inspection Panel's 2002 report on Bujagali, many of the central concerns raised by Ugandan NGOs were confirmed, particularly in relation to the financial analysis of the project, disclosure of information, cumulative environmental impacts, assessment of alternatives and public participation efforts (see Inspection Panel 2002; World Bank 2003):

> The panel found the Bank violated five Operational Procedures and Directives, including OP 4.01, Environmental Assessment, and found the Bank was lacking sound financial and economic analysis of the project and analysis of other power alternatives, such as geothermal. (Linaweaver 2003, p. 292)

Another significant issue the Inspection Panel examined in relation to environmental concerns and the approval process relates to the protection of Kalagala Falls. The 'Request for Inspection' submitted to the World Bank Inspection Panel in 2001 argued that among other violations, the World Bank violated its own Operational Policies on Natural Habitats (OP 4.04). The argument was that Bujagali would seriously harm the tourism industry at Bujagali and that this was not adequately considered in the EIA for Bujagali. In the World Bank Management's Response to the Request for Inspection (which is submitted to the Inspection Panel before the Panel begins its investigation), the Bank acknowledged the dam would inundate Bujagali Falls, that there was no feasible alternative design to limit this impact, and that the impacts on tourism had always been a concern.[3] As a result, the Bank stated that one of the negotiated agreements between the Government of Uganda and the World Bank (formally the International Development Association, IDA) was that another set of falls popular for tourism but also amenable to a hydroelectric dam would not be developed. This agreement was known as the 'Kalagala Offset' – a World Bank requirement to preserve Kalagala 'as an environmental and cultural offset in perpetuity' in order to receive financial guarantees for the completion of Bujagali. In relation to the Operational Policy and concern raised above about the loss of tourism, the World Bank response to the Inspection Panel stated:

> '... as the implementation of the proposed Bujagali Hydropower project will inundate Bujagali Falls, *the World Bank Group concluded that Kalagala Falls must be conserved in perpetuity for its spiritual, natural habitat, environmental, tourism and cultural values*' [emphasis in original] and in paragraph 142, that 'the Kalagala Falls site will be preserved in its present state as per the agreement between the Government of Uganda, IFC and IDA as an environmental off-set. This area is of special interest for local tourism development.' (Letter to V.P. & Gen. Counsel of the World Bank, from the Acting Chair of the Inspection Panel, 14 December 2001)

Hence, in the eyes of the World Bank, it had come to an agreement with the Government of Uganda to preserve Kalagala. The issue, however, was that the Inspection Panel was not sure whether this agreement was legally binding on the Government of Uganda. Hence, in late 2001, the Inspection Panel sought World Bank legal opinion about whether this agreement had legal status. The confusing part of this agreement was that the World Bank Management Response to the Inspection Panel said that an agreement had been reached to offset Kalagala 'in perpetuity'. However, the actual letter signed on this issue (the Mitigation for Loss Agreement) said something different. The formal agreement, as quoted by the Inspection Panel in its letter seeking a legal opinion on this matter, stated: 'The Government of Uganda undertakes that any future proposal which contemplates a hydropower development at Kalagala will be conditional upon a satisfactory EIA being carried out which will meet the World Bank Safeguard Policies as complied with in the Bujagali project. The Government and the World Bank will jointly review and jointly clear such an EIA.' Hence, the Mitigation for Loss Agreement left open the possible development of Kalagala pending a satisfactory EIA.

[3] This information is derived from publicly available documents retrieved from various sources on the Internet, including International Rivers, the Bank Information Centre, the World Bank Inspection Panel, and the World Bank's Project database. As of February 2016, most of the documents can still be found listed here: http://documents.worldbank.org/curated/en/2002/05/3054709/uganda-bujagali-project-inspection-panel-investigation-report. Copies of these communications and letters are held by the author in e-version and can be provided upon request.

Given these contradictory positions, the Inspection Panel asked World Bank Legal Counsel whether the Government of Uganda did in fact agree to preserve Kalagala. The answer was a very clear 'no'. In a response to the Inspection Panel dated 5 March 2002, the letter states that the signed agreements 'do not give rise to a valid, binding, and enforceable obligation of The Republic of Uganda to conserve in perpetuity the Kalagala Falls as an environmental and cultural offset. The lack of any obligation to conserve Kalagala Falls in perpetuity is not inconsistent, however, with OP/BP 4.04 on Natural Habitats.' The World Bank Legal Counsel, in its letter, stated that

The conclusion is based on the following: (a) in the exchange of letters dated April 25, 2001, the only reference to conservation in perpetuity of the Kalagala Falls is contained in a cover letter conveying the proposed Mitigation for Loss Agreement from a World Bank official to a Ugandan official, but not in the actual Mitigation for Loss Agreement whose terms were accepted by The Republic of Uganda; (b) neither the Mitigation for Loss Agreement nor the subsequent Indemnity Agreement contains any provision requiring conservation in perpetuity of the Kalagala Falls; and (c) OP/BP 4.04 does not require conservation in perpetuity.

The details of the situation surrounding Kalagala are important because it revealed that an agreement widely discussed in the media and intended to diminish concerns about economic losses from tourism had no legal standing. The Kalagala offset would arise again when the Bujagali II project was resurrected, discussed in Chapter 5.

Another area the Inspectional Panel examined in relation to Bujagali I was the complexity of the electricity reform process and Bujagali's role in that process. The Inspection Panel wrote that another area of concern

... relates to the privatization and performance of the distribution concession. It is clear that the performance of the distribution sector is likely to play a significant role in the ability of the Bujagali project to deliver sustainable benefits... The distribution sector is key to the connection of new consumers (and so to providing the benefits of access to electricity) and to collecting revenue (and hence to the ability of the power sector to finance its service provision, and to restrain tariff growth to compensate for non-payment). Because of this, the status and performance of the privatized distribution sector is an important element in risk associated with the Project. Correspondingly, therefore, there are some difficult issues: tariffs have to be low enough to be affordable but sufficiently high and sustained to make it worthwhile for a profit-making entity to commit to collecting them. In the Panel's view, an indication of a thorough examination of the institutional risk of a delayed or underperforming privatization of the distribution system, and its impact on the robustness of the Project's affordability is missing from the [Bank's] economic appraisal ... although this was needed for full compliance with [Operational Policy 10.04: Economic Analysis of Projects]. (Inspection Panel 2002, p. xviii)

Reinforcing these findings, in a 2003 interview with Paul Maré, then the Managing Director of Eskom Enterprises and later the General Manager of Umeme, the private company that held the concessions for electricity distribution, Maré said that when project financing for Bujagali became questionable in 2001, 'the concessions were thrown into a loop' (Interview, 17 January 2003). This was the same position of a Nordic donor representative earlier quoted – the privatization process was contingent on Bujagali's successful completion. Owing to the delay in Bujagali's completion, the concession of the distribution network was also delayed and complicated, investments in the entire infrastructure network were delayed, and hopes for improved and increased access to electricity in the near future were severely interrupted.

On 1 March 2005, Umeme officially took over the Uganda Electricity Distribution Company Ltd. Under the company name, Umeme, UK-based Globeleq (owned by the CDC Group) and Eskom Enterprises assumed joint control over UEDCL under a 20-year concession. Eskom Enterprises was the entrepreneurial wing of the state-owned South African electricity company Eskom. In the original arrangement, Globeleq owned 56% of the new company and Eskom Enterprises 44%. It was, however, reported that in 2006 Globeleq assumed 100% of Umeme's shares, with Eskom pulling out of Uganda's distribution service (*East African* 2007). In February 2007, the *East African* also reported that Globeleq was selling its interests in twenty electric power projects in Africa, including Umeme. The Government of Uganda was set to receive over US$350 million from the transaction – the highest return on any privatization concession in the country to that point (*Monitor* 2006). Other financial details of the agreement included a $1.4 million transaction fee; an annual rental fee of $18 million for use of the state distribution company's assets (UEDCL), which no longer distributed electricity but retained ownership of some network assets; and an obligation to invest a minimum of $65 million in the distribution system over five years to upgrade the physical infrastructure, billing system and customer support services. Umeme was also expected to make a minimum of 20,000 annual connections over the next five years.

Umeme took over the distribution system at one of the most troubling points in the country's electricity history. As noted earlier, in 2006 Uganda had one of the highest per unit prices for electricity, and was only able to produce 165 MW of electricity. Electricity distribution losses were routinely near 35% of electricity purchased from the Transmission company owing to poor infrastructure and theft (Ministry of Energy and Minerals Development 2014, p. 11). Adding to this, the country was relying on expensive diesel generators to make up electricity shortfalls owing to unexpectedly low water levels in Lake Victoria, which undermined generation capacity. Umeme had taken over the distribution concession prior to 2006. Hence, financial, ecological and project difficulties converged at a horrible time, creating Uganda's 'electricity crisis', as the Permanent Secretary described it in 2006.

Despite all of this, Umeme's commitment to the original terms of its contract did not change initially. In large part this was because of the Partial Risk Guarantee (PRG) that the World Bank had extended to Umeme – the first ever application of a World Bank PRG to a utility system (see Eberhard 2005): the PRG 'provides support for potential loss of regulated revenues resulting from a "guaranteed event" … These include non-compliance by the regulator of the pre-agreed tariff framework, full pass-through of the bulk electricity tariff supply from UETCL … and timely adjustments of tariffs' (ibid, p. 33). The PRG also addressed non-payment of government agency electricity bills and ensured provisional payments pending dispute resolution during a period of 'contract stress'. According to Anton Eberhard, in a presentation made at the World Bank's 2005 Energy Week, titled 'Good Fences Make Good Neighbours', the CEO of Globeleq (then the majority partner in Umeme) said that the provisional payment feature of the PRG was 'deal-clinching' (Eberhard 2005, p. 33).

In short, the World Bank PRG was tacit recognition of the extraordinary risk involved in the reforms being undertaken and was doing everything it could to salvage and protect the electricity reform process it had championed. But despite the role of the PRG in promoting the much-needed investment, the ongoing electricity supply problem in Uganda eventually led Umeme to reconsider its capacity to achieve the goals originally laid out in its investment. In November 2006, Umeme applied for a review of its operating licence and a restructuring of its concession agreement because the distributor now lacked the electricity supply it had been expecting from the Bujagali dam. Hence, Umeme's potential to achieve the distribution goals established in its concession agreement were going to be severely challenged with a lack of reliable supply. Umeme met its household connection targets by 2008. In fact, Umeme staff acknowledged that they never thought this would be difficult owing to the high demand for electricity and low number of connections. But in 2006, at the lowest point in electricity supply, the company's concern remained very high, with fallout of the problems with the Bujagali I project and dismal supply of electricity making everyone anxious.

The politics of process

The Government of Uganda had never given up on its commitment to construct the Bujagali dam owing to its importance in salvaging the new structure of the electricity market. In early 2004, a second call for tenders to construct the dam – 'Bujagali II' – was issued. One year later, in May 2005, the government announced that the firm Industrial Promotion Services (IPS), a member of the Aga Khan Fund for Economic Development (AKFED) – the economic development arm of the Aga Khan Development Network (AKDN) – along with its partner company, US-based Sithe Global, had successfully outbid five other companies to win the new contract to construct the dam. On the eve of the resurrection of the project, Ugandan officials did not hide their frustration with the previous problems. The critiques were firmly pointed at the World Bank, among others.

At the height of Uganda's electricity crisis, in 2006, former Minister of Energy and Minerals Development, Syda Bbumba, presented a keynote address in the plenary session of the World Bank's Energy Week. In her remarks, Bbumba criticized the Bank's approach in Uganda (Bbumba 2006). Bbumba suggested that in implementing reforms, there must be a recognized transition process, and that resources should be allocated for both market-oriented reforms and public sector delivery efforts. Moreover, based on Uganda's experience trying to establish new distribution concessions in areas of high and low demand, she emphasized that there was no private sector interest in areas of low demand, and therefore the public sector must play a role. Hence, in her view public and private provision of electricity were not mutually exclusive. Indeed, Uganda's neighbours, Kenya and Tanzania, both maintained majority control over their electricity systems while Uganda was undertaking dramatic restructuring, despite having very similar levels of electricity access. In perhaps Bbumba's most critical remarks about the reforms, her presentation slides state:

> As we went about implementing the reforms, it was assumed that we could break away from the traditional public sector delivery and go straight into private delivery models ... Our experience to date has proved this assumption wrong ... The only conclusion that can be drawn is, therefore, that there is a need to re-examine and redesign the strategies and the programmes that we have put in place with the help of our development partners, basing them on the realities of each reforming country other than the 'one-size-fits-all' prescription, which, apparently, is now being applied.

Striking as these remarks were, the problems and challenges in Uganda were not unnoticed within the Bank. The Bank's own evaluations in the early 2000s, at the height of controversy surrounding the first Bujagali project, pointed to critical procedural concerns and coordination problems between Washington and resident staff. In a 2001 Operations Evaluation Department report titled 'Policy, Participation, People', the authors wrote that 'against the framework of [an] impressive list of achievements and strengths' the Bank also has weaknesses 'switching from macroeconomic to sector and thematic reform' and

... that IDA's project implementation suffers from poor design and sequencing, rigid and confusing procedures (particularly for procurement and disbursement), frequent changes in task managers, injudicious reliance on project implementation units, and poor monitoring and evaluation ... The resident mission lacks the requisite procurement and sectoral expertise and decision making power because task managers in Washington generally make decisions. (World Bank 2001b, p. 40)

In short, the Bank's general evaluation stated clearly the difficulty of moving from first- to second-generation reform procedures and the difficulty of implementing complex reforms.

In 2006, the Nordic Consulting Group (NCG) also evaluated the bilateral agency's energy lending on behalf of the Norwegian Agency for Development Cooperation (NORAD). Since the early 1990s, NORAD, along with GTZ, had been one of the principal bilateral donors working on energy issues in Uganda. Between 1997 and 2005, NORAD contributed approximately US$54 million to 25 electricity projects (Nordic Consulting Group 2006, p. 1). One of the things discussed in the NCG report was the impact of the World Bank's reform strategies in Uganda. Its assessment is critical and speaks directly to the problems that followed from Bujagali I's initial failure, and the risk associated with linking state sector reform to the dam. In its overall assessment, NCG first noted the ongoing problems in Uganda's energy sector. It then spoke directly to the problems the Bujagali I project had brought to the sector overall. Here are some of the report's key findings:

1) The current power crisis has led to a dramatic reduction in production capacity and an increasing gap between demand and supply. The target of electrifying 10% of the rural population by 2010 will most likely not be achieved.[4]
2) While the aim of the reform process was to promote a commercially viable sector with limited requirements for state subsidies, the current situation is the opposite. Recent estimates suggest that the GoU will be required to provide direct and indirect subsidies to the tune of USD 420 million over the next 4 years to support a tariff below prohibitive levels for the consumers.

---

[4] Based on the number of households connected to Umeme's electricity network in 2010 (1.3 million) and an average national household size of 5 (based on Uganda Bureau of Statistics data for 2009/10) and an estimated population in 2010 of 30.7 million, approximately 20% of the national population had access to grid-based electricity in 2010. This figure might be slightly higher owing to connections to independent networks, but there were very few at the time.

3) While technical efficiency has improved, overall efficiency varies significantly from one year to another without a clear trend and total system losses remain high (35–40%). The private sector response to the new regulatory environment has so far been very limited with few projects considered by even fewer potential investors. Delays in negotiations over large-scale investment projects like Bujagali for which the GoU and donor/IFIs have shown a particular preference, may partly serve to explain the limited response by the private sector for other investments.

4) Numerous sector studies in the 1990's pointed to the fact that Uganda has a largely untapped hydropower potential which could generate significant export revenue for the country. However, Uganda is now in a situation where it has to invest in high cost thermal power to compensate for some of the domestic supply losses.

5) The GoU now faces a situation in which few of its targets for the sector reform will be met, and the power sector will demand a record high share of scarce public funds initially intended for other priority expenditure. It has led to a reduction rather than increase in access to power and has had adverse impacts on rural access, contrary to its strategy.

6) With the commissioning of Bujagali a considerable surplus of electricity was expected and the Government was reluctant to enter into other Power Purchase Agreements. *Instead of spreading the risk, a lot was 'put into one basket', and when this did not come out as expected, there was little to fall back on.* (NCG 2006, pp. 1–2, emphasis added)

This assessment, combined with the frustrations of Uganda's former Energy Minister, clearly point out the importance of the process, sound reform design and implementation, and, ultimately, the principles underlying the procedural character of energy transitions. The World Bank's own assessment of the challenges it faced in sector reform in Uganda, combined with NGO frustrations with the absence of substantive debate in reform, further illuminated the importance of treating all the conditions deemed necessary for successful reform carefully and substantively. But together, these assessments highlight a much simpler point: Uganda's electricity sector reform plan was enormously complex, ambitious and risky. Multiple transformations were expected to take place simultaneously: technical transformations in the infrastructure; financial transformations surrounding electricity pricing and project financing; bureaucratic transformations with respect to the new roles of several large bureaucracies focused on energy issues; and attempts at political transformations by using reform to change the relationship between elected elite and their influence over service provision.

One of the major problems was the lack of capacity in the Ministry of Energy to help guide this change. As one European donor representative told me in 2002, the Ministry of Energy was tasked with overseeing reforms and the construction of the Bujagali dam, but 'its absorptive capacity [was] quite limited' (Interview, 18 March 2002). It was severely overwhelmed by the multiple reforms taking place. The capacity of individuals in the Ministry was high, but the small staff was completely overwhelmed by donor and government requirements and expectations. In 2002, the electricity division of the Ministry of Energy had only three staff. Despite this, during the revision to the Electricity Act, the Commissioner responsible for electricity worked non-stop on the Act for two months, but was routinely interrupted by calls from Members of Parliament wanting information on their bills or offering suggestions. In addition to domestic demands, the World Bank was reported to have five separate energy teams visiting Uganda during the lead up to and during the height of the reform and dam construction effort. In some cases, teams of fifteen World Bank staff or consultants were visiting Uganda every two months, requiring time and information from the Ministry staff. In short, the reform agenda placed enormous strain on key government departments with limited capacity and experience to execute them.

In addition to unbundling and reforming the sector, and trying to construct Bujagali I, the Energy for Rural Transformation I project, which was intended to expand access to electricity for rural residents, was not nearly as successful as anticipated. Indeed, this project was initiated in the same year that the Rural Electrification Agency was created, which is also the same year that the new Electricity Act was passed, 1999.

Another transformation resulting from the reform experience, which was not purposeful but was significant, was a social transformation: civil society organizations were using existing state structures and international institutions to challenge state goals, and they were challenging the state in new ways, making sophisticated arguments about reform processes and outcomes, as well as about transparency and accountability. Most critical of all was the social impact: the focus on the Bujagali dam as a response to Uganda's electricity supply problem meant that all other electricity-related initiatives took a back seat and household access to electricity failed to improve significantly.

In 2008, I returned to Uganda for the first time since 2003. I requested a meeting with a World Bank official deeply engaged with the electricity sector. The staff person accepted to meet me. But the environment for the meeting was one of the most tense I had ever encountered. We did not meet in an office but in a boardroom. The staff member requested to record our conversation, which I accepted. I asked how the Bank's approach to electricity in Uganda had changed since Bujagali I and what the Bank had learned. The Bank staff member paused and said: 'Why don't you tell me how you think the Bank's approach has changed and what it should have learned?' Despite this sharp retort, I calmly responded to explain and note the things that were problematic and that I identified above: the ambition of the effort and link to the dam, the financial and political risk, and the unexpected role and challenges of civil society. The Bank staff member's edge eventually decreased. The key moment in this discussion was a subtle admission that the reform agenda had been too risky.

A chance encounter three years later with another World Bank staff member who had been working in Uganda at the height of the Bujagali I project echoed this. While at the World Bank Tanzania office in June 2011 I was introduced to someone who had worked on the electricity reforms in Uganda. When I asked him about Uganda's experience, the staff member laughed and said, in reference to Uganda: 'You know how Tanzanians are sometimes considered to be more careful and cautious than their East African neighbours? Well, sometimes it pays to be cautious!' He went on to say that the slow introduction of reforms and independent power producers in Tanzania would have been a better approach for Uganda. These admissions were striking but also deeply troubling given that Uganda's experience with Bujagali I meant that the country continued to suffer with poor access to electricity for several years. But the experience is also very significant for it fundamentally altered the country's approach to electricity and its future relationship with the World Bank on matters of electricity.

5
Electricity
& the Politics
of Transformation

The Government of Uganda and the World Bank's commitment to the Bujagali project did not wane. In April 2007, the World Bank Group approved US$360 million in loans and guarantees for the project (Bujagali II) – $130 million in loans to Bujagali Energy Ltd (BEL) from the International Finance Corporation; a Partial Risk Guarantee of up to $115 million from the International Development Association; and an investment guarantee of up to $115 million from the Multilateral Investment Guarantee Agency. Financial support for the project also came from the African Development Bank ($110 million) along with the European Investment Bank and the German Bank for Development. The new total estimated cost of the project was US$799 million. Up almost US$300 million from the original project cost, Ministry of Energy sources reported that the increased cost was a result of higher prices for oil, cement, steel, iron and consultancy services (Mugirya 2007).

The second Bujagali project (Bujagali II) did not go without controversy, however. The National Association of Professional Environmentalists (NAPE) continued to voice concerns about the project. Their major anxieties remained the cost of the project, the expected rise in future electricity tariffs, and hydrological concerns surrounding drought and climate, fisheries and protected land. Moreover, NAPE continued to take its concerns to the World Bank and other project financiers. Senior Bank officials responded to these concerns openly and directly. Michel Wormser, World Bank Sector Director for Sustainable Development, Africa Region, stated: 'The World Bank Management remains committed to the successful implementation of this project including the appropriate application of relevant environmental and social safeguards ... The project is critical to Uganda's economic development and we will continue to work with the Government to ensure that this project meets high standards' (Kasita 2007). In the same interview with Ugandan media, Wormser said: 'The project's approval reflected a shared view by management and the board of the critical importance of providing a new source of electricity expeditiously to Uganda and confidence that thorough economic, environmental and social due diligence has been undertaken to identify and realise that source' (ibid). As one indicator that the Bank learned from some of the transparency problems encountered in its support for the dam, it created a comprehensive website solely dedicated to the project. Bujagali Energy Ltd, the private project sponsor, also established its own comprehensive website, still functioning as of March 2017 (www.bujagali-energy.com). It is noteworthy that in the 'About Us' section of the website it stated: 'Bujagali Energy Ltd is not associated with AES Nile Power Ltd (AESNP), the previous sponsor of a similar proposed project in Uganda.'

A second way that the Bank tried to assert its commitment to environmental and social impacts was to link project support for Bujagali II to an Indemnity Agreement. The Indemnity Agreement specifically referred to the Kalagala Offset, which the Bank had mistakenly thought was legally protected when it approved the Bujagali I project (as described in Chapter 4). The Indemnity Agreement stated:

> Uganda shall set aside the Kalagala Falls Site exclusively to protect its natural habitat and environmental and spiritual values in conformity with sound social and environmental standards acceptable to the Association. Any tourism development at the Kalagala Falls Site will be carried out only in a manner acceptable to the Association and in accordance with the aforementioned standards. Uganda also agrees that it will not develop power generation that could adversely affect the ability to maintain the above-stated protection at the Kalagala Falls Site without the prior agreement of the Association. (Section 3.06, Indemnity Agreement, 18 July 2007)

Hence, the Bank corrected the legal error made in the original agreement for Bujagali I. This agreement is significant for clearly attempting to restrict Uganda's autonomy over the use of its own resources in future. Further, it is not clear what consequences would arise if Uganda disregarded this agreement in future. This matter is important because the Kalagala Offset arose again in 2013 when construction began on a new dam, this time financed by China (described further below).

In early September 2007, the physical construction of the Bujagali dam had begun. It was completed in February 2012, bringing 50 MW of power initially to the national grid, with 200 MW more in the months that followed. While Uganda would continue to suffer electricity supply shortfalls for months following the dam's inauguration, eventually the 250 MW project eliminated most of the country's reliance on costly diesel generators. Nonetheless, many challenges persisted in the months leading up to and after the project's completion. Two central concerns were the cost of electricity and the independence of the Electricity Regulatory Authority (ERA).

One month before Bujagali II was complete in February 2012, the ERA communicated that a large increase in electricity tariffs would be instituted. There were two key reasons for this. In 2010, ERA originally proposed a large tariff increase. Consultations with the public and private sector, not surprisingly, revealed strong opposition to an increase. Members of Parliament (MPs) also expressed vigorous disagreement with an increase, and the ERA withdrew the suggested tariff revision. However, in January 2012, a month prior to the inauguration of Bujagali II, a new price increase was advertised. Again, MPs protested strenuously, with one stating: 'We demand that you withdraw these rates or else we shall send the people onto the streets to demonstrate against you' (Talemwa 2012). The anger was not surprising given that tariffs were expected to rise by nearly 70%. But the rationale for the increase stemmed from three facts: (1) the simple need for the generation, transmission and distribution companies to raise revenue to invest in the system; (2) the previous delay in increasing tariffs; and (3) that the Government of Uganda was going to end the subsidy it had been paying since 2005 to keep the price of electricity down.

Since 2005, the Government of Uganda had been paying US$200 million annually to keep electricity tariffs lower than the unsubsidized rate. ERA also declared publicly that from April 2012 forward, tariffs would be reviewed monthly and that the public should expect the price of electricity to continue to increase. In response to the question about whether the power from Bujagali would decrease the price of electricity, ERA Executive Director Mutambi stated: 'So far there is nothing to show that once Bujagali comes on stream, the cost of power will come down. So, it is unrealistic to expect lower power tariffs' (Talemwa 2012). This was a striking admission and shocking to MPs who had always hoped and communicated that Bujagali's commissioning would reduce tariffs. Demonstrating the disconnect between narratives that circulated about the benefits of the Bujagali project and reality, many analyses of Bujagali had shown that the price of electricity would rise for many years after being commissioned. Indeed, in the early 2000s, bilateral donor representatives confidentially showed me reports that clearly indicated that price of electricity would continue to increase in Uganda after Bujagali came online. These, however, were not shared publicly or in the mainstream. Another thing that the 2012 price increase revealed was how the new regulatory system would function in practice and the tensions that would result from greater regulatory independence.

When ERA announced the high increase in tariffs, the board was summoned to Parliament by an ad hoc energy committee. MPs claimed that the increase was illegal and demanded that ERA halt the price increase. Of course, the increases were not illegal. ERA pointed out that the Electricity Act provided no role for Parliament in setting or reviewing tariffs and that ERA had sole authority. The conflict with ERA's independence was not new. As far back as 2003, the then Exec-

utive Director of ERA, Dr Frank Ssebowa, referred to political interference in the authority of the regulator (Eremu 2003). Ssebowa even
challenged the government's policy on subsidizing the price of electricity on the basis that only 5% of the population had electricity:

> We have to convince Ugandans that electricity is an expensive
> commodity. It is not a social service. A social service should affect all
> Ugandans equally. But we are talking of only 5%. It is the most unfair
> situation I have ever heard of in my life. My suggestion to Ugandans
> is to be realistic. In my view, the 5% who have electricity are unfair
> to the 95% if they insist on being subsidised even further. (ibid)

My interview with Dr Ssebowa in 2008 revealed that he was not
against subsidizing the cost of connecting Ugandans to the distribution
network; he was just against subsidizing the price of electricity. Hence,
implicitly, regulators were engaged with questions about distributional
equity and justice.

These tensions around political independence and subsidies illustrate how electricity reform and dam construction were spawning
several transformations in the country: not only was there a clear goal
to transform the electricity system and access to electricity but to also
challenge and alter the political environment through institutional and
regulatory mechanisms. This most certainly was also an uncommuni-
cated goal of the World Bank in promoting these reforms; it aimed to
limit political interference in the regulation of the electricity market.

In 2016, conflict and tension between different government institutions responsible for electricity remained prominent. There remained a
lack of clarity about who owned state assets, for example, which led to a
review and revision of the 2001 Electricity Act. Further, concerns about
the quality of the work on two new dams being built revealed tension
between the Uganda Electricity Generation Company Ltd (UEGCL) and
the Ministry of Energy. The conflict related to who had control and
oversight over the projects and the potential financial benefits that
may accrue to the organizations from that oversight (see Matsiko 2016).
The other fundamental change that materialized after the first failed
effort to construct Bujagali was a change in how Uganda was going to
fund future electricity projects and who it would partner with in those
endeavours.

Learning by doing

In a report reviewing World Bank urban projects from 1972 to 1982,
the authors titled the study 'Learning by Doing'. In the report, Cohen
et al. wrote, 'In 1972, given the lack of solutions to urban problems,
coupling learning with doing was the only sensible approach the Bank
could take as it entered a new sector of lending' (1983, p. 2). This mantra

has not disappeared from the Bank's work (see World Bank 2013), and of course applying lessons from the experience of 'doing' is important in policy implementation and assessment. But in relation to Uganda's experience with electricity and hydroelectric projects, it seems that the lessons Uganda learned through the World Bank's 'doing' were much more profound and dramatic than was probably anticipated. As I noted earlier, the World Bank certainly learned from its experience in Uganda as well. For example, it codified the offset area for Kalagala Falls in Bujagali II after doing so incorrectly in Bujagali I; it slowed or reduced the ambition of the reform agenda in other countries, including Tanzania and Kenya; and it did not promote privatization in other Ugandan sectors like water after its experience with electricity owing to the challenges it encountered.

But beyond the World Bank, the Government of Uganda was also 'learning'. By 'doing' electricity reform and dam construction in the manner the Bank advised, the GoU also took away some significant lessons that transformed its relations with donors and approach to the electricity sector. Two key lessons were that it did not want to rely on the World Bank for financial support for future hydroelectric schemes and, more broadly, that it would not rely on its customary development partners to fund future, large, hydroelectric schemes.

Around the same time that the country faced its most critical electricity shortfall (2006) and the Bujagali II project was resurrected (2007), the GoU began to save approximately US$70 million per year to finance future hydroelectric schemes. This money was raised by adding a surcharge to the bills of electricity consumers. The rationale for taking this action was clear, as was the anger and frustration towards the World Bank.

In a 2008 speech on investment, President Museveni attacked the World Bank, arguing it had delayed or stopped hydroelectric development and caused power shortages (Muwanga 2008). He then went on to state that with the new Energy Fund, the country could address the problem of relying on the World Bank: 'I no longer spend sleepless nights [worrying] about people coming to build dams in Uganda. If they come, they are welcome. But if they don't, we shall do it ourselves.' Part of this frustration also stemmed from the fact that in 2008 Norpak Power pulled out of its effort to construct the 600 MW Karuma dam. Once again, there are conflicting reports about the reason for Norpak's withdrawal. Norpak claimed it was a consequence of ongoing protracted disputes with the World Bank. It is not clear whether the rules imbedded in the original PPA for Bujagali I remained in place for Bujagali II, which required Bujagali to be complete before Karuma could begin. Meanwhile, the Electricity Regulatory Authority argued that Norpak's withdrawal was due to a financial crunch and failure to raise a performance bond of US$300,000 as a commitment to implement the Karuma project (Kasita 2008). Whatever the explanation, in 2008, the intellectual property of the Karuma project was turned over to the Government of Uganda and it sought out new partners, particularly in China.

China, dams and donor realignment
Uganda and China have held formal diplomatic relations for over fifty years. The President of Uganda has travelled to China on several occasions encouraging greater South–South cooperation. Chinese support for infrastructure development in Uganda has also been steadily increasing in recent years, moving from roads, hospitals and railways to communications infrastructure, government ministry buildings such as the Ministry of Foreign Affairs and Presidential Office, and the national stadium (Jaramogi 2014). Uganda also clearly recognized the high potential for Chinese support for hydroelectric development. In fact, in 2006, at the height of Uganda's electricity woes, Uganda's Energy Minister attended the Sino-Africa summit in Beijing seeking investors for the sector.

The year 2013 turned out to be a monumental one for Uganda–Chinese hydroelectricity relations: Sinohydro Corporation Ltd won the contract to build the Karuma project; the China International Water & Electric Corporation (CWEC) won the contract to build the 183 MW Isimba dam; and the Gezhouba Group won the contract to build the 600 MW Ayago dam. Sinohydro is a Chinese state-owned company; the Gezhouba Group is partially state-owned by the China Energy Engineering Corporation or Energy China; CWEC is owned by the state-owned company China Three Gorges Corporation. These dams are being financed primarily with loans from the Export-Import Bank of China (Exim) and with funds from Uganda's Energy Fund, which is often paying for the transmission lines for the dams.

Having dealt with the World Bank extensively throughout its electricity sector reform experience, and relying on the Bank Group and European Export Development Corporations for project financing for the Bujagali project, Uganda looked elsewhere for development partners. Hence, the experience trying to build the Bujagali dam had a profound effect on who Uganda wished to partner with in future large electricity generation projects: the country turned away from the World Bank and European donors to finance future large hydroelectric projects.

In 2012, a representative from KfW Entwicklungsbank, a European lender, stated: 'Various development partners have made offers to support the implementation of the Karuma Project, for instance through technical advisers or financing for an international panel of experts for dam safety. Thus far, these offers have not been taken up' (Nakkazi 2012). Nakkazi (2012) reported that Ministry of Energy officials wanted to avoid the 'kind of environmental and financial noise that frustrated the first attempt at building the Bujagali power station', and were unapologetic about their stance. Bukenya Matovu, head of communications at the Ministry of Energy and Mineral Development, said: 'Having learnt a lesson from Bujagali, we are not prepared to go through that again' (Nakkazi 2012). But the absence of World Bank involvement also means that Ugandans are also stuck trying to independently resolve problems when they arise, as was the case in 2016 when cracks in the concrete of two of the three dams being built by Chinese firms were discovered. The result, as noted earlier, was tension between various Ugandan electricity agencies, infighting, finger-pointing and deference to the President to resolve the conflicts and mess (Okuda 2016).

Another significant outcome has been a realignment or shift in donor–state and donor–donor relations, and an implicit recognition that large dam projects will be led by Chinese firms and Chinese finance. This situation has placed the World Bank in a new role in Uganda as well. In 2017, there was an unresolved tension over the potential impact of a Chinese-financed dam on the ecological area deemed protected by the World Bank – the Kalagala Falls Offset, discussed earlier. One of Uganda's new dams, the Isimba dam, is a large (183 MW) hydroelectric project on the Nile, 40 km north of the Bujagali dam. Based on the original designs of the Isimba dam, a large area that is part of the Kalagala Offset will likely be flooded and altered by the Isimba reservoir, with some falls flooded and natural habitats affected. As a result, the World Bank has been arguing that safeguards must be put in place to protect the Kalagala area from negative environmental and social impacts, and that the Government of Uganda must adhere to the 2007 Bujagali Indemnity Agreement, which is supposed to protect the Kalagala Falls.

The Isimba dam, however, is not being built with World Bank funds: the dam is being constructed by the China Water & Electric Corporation (CWEC) and is receiving financing from the China Export-Import Bank (Exim). CWEC is also constructing a transmission line between Isimba and Bujagali. Hence, the World Bank is now arguing that the Government of Uganda must comply with the 2007 Bujagali (II) Kalagala Offset, but in relation to the impacts of a dam project in which it is not involved. World Bank media releases (World Bank 2015b) and news reports (Musisi 2017) claim that the GoU must comply with the Indemnity Agreement, but it is not clear what sanctions the Bank would apply if it deems that the conditions are violated. Further, what actions would China take if the World Bank tried to sanction a GoU project being built by a Chinese state-owned company and financed by the Chinese state? These are very new questions with little precedent.

One consistent concern raised about Chinese involvement in Africa is that it disregards environmental and social impacts. The reality, not surprisingly, is much more complex. For example, the Exim Bank is a signatory to the Equator Principles – a risk management framework adopted by financial institutions to assess and manage environmental and social risks of projects. The Exim Bank has also published its own guidelines for social and environmental assessment, which include reference to land rights and resettlement concerns (Braütigam 2011, p. 121). But, as Bräutigam notes, 'there can be a wide gap between guidelines and actual project funding', and 'Without considerable more transparency, it will be difficult to know the extent to which these guidelines are actually applied by China' (ibid).

Of course, Uganda also has its own Environmental and Social Impact Assessment policies, which govern the Isimba dam project. The Uganda Electricity Generation Company Ltd (UEGCL) oversees all new dam projects in Uganda. It has countered claims about negative social and environmental impacts around the Kalagala site: 'Kalagala Falls will not be affected or submerged completely as feared. And so it is not entirely true to posit that there will be a total loss of the rapids and livelihood as a result,' stated Simon Kaysate, spokesperson for UEGCL (Musisi 2017). Yet at the same time, UEGCL also stated that losses from tourism at Kalagala are outweighed by the benefits of hydropower and alternative tourism opportunities. The spokesperson claimed that an addendum to the Kalagala offset has been prepared to address environmental and social impacts of the Isimba project, so time will tell what materializes. At the time of writing, this was still not resolved. Nonetheless, it is striking that a debate about a loss of tourism opportunities versus hydropower have emerged once again in Uganda – the first time shortly after the end of the colonial era in relation to Murchison Falls; then for Bujagali Falls; and now for the Kalagala Falls.

These events and China's involvement do not mean that customary development partners are not involved in Uganda's electricity sector or have retreated from it. They remain active but in a very specific way. In Ethiopia (2010), Nordic donor representatives explained to me that their new role was in capacity-building and transmission, distribution and small renewables. This was because they were not invited to participate in large dam projects being built in the country. Following the slow, complicated and conflict-laden processes surrounding the first efforts to build the Bujagali and Karuma dams in Uganda, a similar scenario seems to have emerged in Uganda. There now appears to be an implicit if not explicit divide in roles: China is supporting large, controversial hydroelectric schemes; European and other customary bilateral donors are focused on small renewables, mini-grids, grid extension and capacity-building.

One of the most exciting examples of European engagement in Uganda's electricity sector was the feed-in-tariff programme, the Global Energy Transfer Feed-in-Tariff (GET FiT) Program, launched in 2013. The programme focused on promoting and installing private, small-scale renewable energy generation projects. Uganda already had a Renewable Feed-in-Tariff (REFiT), but the new programme aimed to make private sector investment in renewable electricity projects more financially viable. The goal of the programme was to foster private investment in renewable generation projects, reduce carbon dioxide emissions, and fill the deficit in electricity by diversifying the energy supply mix. The first phase of the programme saw several solar photovoltaic projects initiated.

The list of key stakeholders supporting the GET FiT programme reveals high European involvement: Norway; UK Aid; UK Department of Energy & Climate Change; German Cooperation; KfW Entwicklungs-bank (part of the KfW Banking Group); the European Union; the Deutsche Bank Group; and the World Bank. The World Bank was offering Partial Risk Guarantees (PRGs) for investors, and the Multilateral Investment Guarantee Agency (MIGA) offered political risk insurance. GET FiT estimates that the programme has leveraged close to US$400 million in private investments; will have facilitated and/or improved access to electricity for 1.2 million Ugandans, and will remove 11 million tons of CO_2 over the course of the 20-year lifespan of the power purchase agreements (GET FiT 2015, p. 9). When the programme ends, it is expecting to have added nearly 170 MW of new electricity generation in Uganda. The programme has attracted a high number of private investors, with Uganda ranked as the second 'most compelling markets for renewable energy investment' in 2015 in the Fieldstone Africa Renewables Index, which ranks countries on their suitability for investment to achieve successful renewable projects (Fieldstone Africa 2015).

Thus, it is clear that there is a great deal of ongoing support and interest in Uganda's electricity sector from a range of bilateral and multilateral actors. What is also very noteworthy is that despite bilat-eral donors having high anxiety over the way that liberalization and reform was done in Uganda, one of the outcomes of the liberalization is that opportunity was made available for private sector-led renewable energy contracts in the country. The first step in Uganda's electricity sector reform process was to amend the Electricity Act to permit the entry of independent power producers. As a result of that, the regula-tory and institutional structure was laid for European bilateral agencies and multilateral agencies to take on a new role in supporting private sector-led renewable electricity projects in Uganda.

One of the central intents of GET FiT was to fast-track small-scale renewable electricity projects, particularly to help satisfy electricity demand while Uganda waited for its many hydroelectric dams to come on line. But the GET FiT programme was temporary, with no new calls for projects forthcoming and all projects expecting to close by 2018. Further, it was intended as an experiment, with plans in 2017 to replicate the model in other African countries owing to its perceived success. Hence, it is interesting that Uganda once again finds itself as a host for an experiment in electricity market reform. In the case of GET FiT it was an experiment that appears to be working well with respect to quickly adding new, sustainable electricity generation capacity. But for Ugandan electricity planners and managers, the legacy of the GET FiT programme still presents challenges.

One of the significant difficulties with these decentralized mini-grids is that it is very hard to predict social response to new electricity supplies. Ugandan energy planners very familiar with the GET FiT programme explained that in some cases demand has been so high that there is no longer sufficient supply. When this happens the Rural Electrification Agency does try to rapidly expand the grid to these areas, but then there is the potential for differential pricing between connections to the main grid and the mini-grid in the same area. In other cases, demand has been very low, yet the private firm has a guaranteed financial return; and in another case, the price of electricity was too high for residents so they chose to disconnect, preferring to wait for the main grid to arrive, which was unlikely for the foreseeable future (Interview, Ugandan energy planners, 21 April 2016). Thus, even amidst exciting innovations in electricity provision and expansion, some very difficult questions about equitable access to electricity and what the future of energy governance in the country will look like remain unanswered.

Energy poverty, justice and governance

Goal 7 of the Sustainable Development Goals (SDGs) is to 'Ensure access to affordable, reliable, sustainable and modern energy for all.' This goal and the accompanying sub-goals are expected to be achieved by 2030. In the words of the World Bank Independent Evaluation Group,

> This is a daunting challenge: more than 1 billion people do not have access, and another 1 billion have chronically inadequate or unreliable service. Most of these without access are poor, and the largest share is in Sub-Saharan Africa. (Independent Evaluation Group 2015, p. xiii)

The report 'Highlights' goes on to emphasize the extraordinary financial investment needed:

> Achieving universal access within 15 years for low-access countries (those with under 50 percent coverage) requires a quantum leap from their present pace of 1.6 million connections per year to 14.6 million per year until 2030. The investment needed would be about $37 billion per year, including erasing generation deficits and meeting demand from economic growth. By comparison, in recent years, low-access countries received an average of $3.6 billion per year for their electricity sectors from public and private sources, including $1.5 billion from the World Bank Group. (ibid)

The report summarizes that of the 1.1 billion people without electricity access in the world, 591 million are in sub-Saharan Africa; the region accounts for 40 of the world's 51 countries with access less than 50% of the population; 22 countries in the region have less than 25%

access; and 7 of these countries have less than 10% access (Independent Evaluation Group 2015, p. xiv). Today, in Uganda, according to International Energy Agency statistics and World Bank World Development Indicators figures, the national household access rate is estimated to be near 18–20% of the national population. Given these figures, there are few words to describe the scale of the challenge facing governments as they choose from innumerable different pathways of intervention and/or are encouraged or required to implement disruptive interventions, such as those in Uganda.

The Government of Uganda did not originally intend to divest from electricity distribution and generation. When widespread divestment from public enterprises began, electricity was designated as a public enterprise to be retained. Over a short time, however, the decision to dramatically reform the sector was taken. It was clear that the Uganda Electricity Board (UEB) had serious service delivery problems. While many of these problems were internal, the external political environment was also recognized to be highly problematic. Senior government leaders deemed their access to electricity a right, and MPs regularly told their constituents that they should and would receive electricity. Owing to these challenges, and the World Bank's mounting frustration with a lack of improvement in the electricity sector despite many reform projects, the Bank decided that no more support would be forthcoming without deep, structural change to the sector. The Bank had no confidence in the state utility. The UEB's problems and the generation needs of the country therefore became an impetus for an ambitious, complex reform agenda, which brought together the unbundling of the state monopoly, the desire to construct a new large hydroelectric dam, new regulatory oversight and the participation of private firms. It was hoped or envisaged that these dramatic, risky reforms would be transformational. The reforms were indeed transformational, but in several unintended ways.

Reducing energy poverty and 'providing energy for all' may have been long-term goals communicated as rationales for Uganda's ambitious reform agenda, but as was shown in this book, how and when these goals might be achieved were debated and contested. At the time of writing, debate continues over what 'modern energy access' means in practice and whether 'energy justice' is achievable in the near term in sub-Saharan Africa. One reason these concepts do not factor centrally into reform debates is because they are conflict-laden and produce deep ethical conundrums for energy planners, donors and politicians. Can electricity be provided in a quantity and quality that is equitable so that individuals, firms and households have equal capability to enhance their well-being? Further, who gets to make these decisions in sub-Saharan Africa? What opportunities exist to deliberate the realities of energy poverty, and do these deliberations have the potential to alter the energy pathways chosen?

These procedural and distributional dimensions of energy justice must be considered in future, not only because they have very real impacts on technical decisions relating to electricity systems and planning, but because *not* reflecting on them can produce very tangible, indirect social and political effects, as Hirschman noted five decades ago: the benefits of a cooperative process may be difficult to evaluate, but their absence may inflict penalties '*that are anything but nebulous*' (Hirschman 1967, p. 163, emphasis in original).

In Uganda, the problems with the electricity reform process, in particular tying unbundling to one large dam project, in the short term, resulted in decreased economic growth, very costly remedies to fill electricity supply gaps, and an inability to meet electricity demand. Uganda still has one of the lowest levels of access to electricity in the world. These tangible 'penalties' are matched by intangible 'penalties' that continue to be felt.

Arguments for more inclusive, deliberative or transparent processes cannot rest on superficial and problematic statements that more participation or openness is better. A misreading of the concerns and capacity of civil society in Uganda, for example, resulted in project delays as civil society successfully challenged the World Bank and government's agenda in international forums (the Inspection Panel) and domestically, using the courts and Parliament. While many would argue that civil society in Uganda was just self-interested and a conduit for international advocates, my research reveals this was not the case – civil society organizations were asking and continue to ask important and sophisticated questions about project goals, outcomes, impacts and beneficiaries. Reforms in Uganda were also being proposed in a political environment where civil society was more emboldened and had more capacity to challenge government policy pathways. Reforms were also taking place in an institutional landscape that offered formal domestic and international mechanisms to challenge the state on poor process if and when it materialized. Hence, the Government and the World Bank did not anticipate civil society's desire and capacity to examine and engage in the reform process; did not heed the political transformation that had materialized in the country; did not anticipate how the evolving political landscape would affect the reforms; and, crucially, did not consider how problems with the reform path would influence future state–society, state–donor and donor–donor relations.

Some twenty-five years ago, James Ferguson (1992) argued that it is incumbent on policymakers and researchers to understand and document the social and political conditions that perpetuate projects and paths of reform, and to recognize the political and social outcomes these paths produce. Building on the political economy approach to the study of energy transitions discussed in Chapter 1, this book has prioritized the need to know more about how governments are situated in these transitions and to recognize that electricity sectors are not just in transition, but are being fundamentally transformed as a result of technical, social, political and economic forces both within and beyond their control. Ultimately, it is individual or collective groups of actors that make decisions to promote and institute different pathways, and therefore, who participates in and controls processes of decision-making is central to understanding past and future pathways chosen. Governments are just one actor among many who may be influential in choosing pathways. And as we have learned in Uganda, the strength or autonomy of governments does evolve over time. For this reason, understanding the evolution and character of relations between actors in the African electricity sectors – understanding the character of African energy governance – is critical for understanding how energy transformations materialize in practice.

Using the notion of 'governance' to study electricity in Africa recognizes that governments are but one actor among many that shape electricity outcomes. The notion of governance draws attention to the relationship between *actors* in a particular process, the *knowledge* that is included and excluded from a process, the *spaces* or opportunities for deliberation or engagement provided, and the *institutions* or rules that condition and structure interactions. A focus on 'energy governance' in Africa, therefore, offers a lens through which we can view the multilevel power relations in a given process, how certain ideas and approaches come to dominate at the exclusion of others, and how different actors shape energy transitions and transformations. In the case of Uganda, the lens of 'energy governance' showed what role the national government has played in the energy pathways of the country, which institutions were important, the circumstances when the government had little influence compared to other actors, and its role in shaping the opportunities to permit and deny deliberation. Hence, energy governance in the country has changed and is still changing rapidly. Uganda reveals how and why the multilevel character of relations in an energy poor country are changing and signals the need to examine deeply, through qualitative research, how and why these relationships are changing. Future research that uses mixed methods to understand the shift in power relations between actors and the impact of those relationships on access to electricity is needed.

This book has also revealed that sub-Saharan African countries, and Uganda in particular, are in the midst of profound experiments in electricity provision, whereby different models of provision are being applied and tested in historically unprecedented ways. These experiments are occurring in the midst of and also as a result of tremendous political change. Hence, as the character of energy governance in African countries evolves quickly and in some cases dramatically, it is incumbent on researchers to examine the short-, medium- and long-term impacts of energy governance on electricity access. Moreover, if 'sustainable energy for all' is truly a goal for Africa, then transition processes must reconcile the relationship between technical and political tensions embedded in these transitions.

African civil society organizations and citizens will continue to advocate for mechanisms and processes that produce equitable and distributionally just energy outcomes. The technical requirements to produce these outcomes are critical. But if energy justice is to be realized, the procedural dimension of energy justice must not be lost. Electricity and electricity reform is deeply political. Researchers, energy planners and development project proponents must acknowledge that technical innovations and market and regulatory reforms function in multiple political time frames: they are conditioned by present processes and will also condition the future processes and political environments that electricity pathways, transitions and transformations are embedded in and dependent upon.

BIBLIOGRAPHY

Advocates Coalition for Environment and Development (ACODE). (2002). *Consolidating Environmental Democracy in Uganda Through Access to Justice, Information and Participation.* Kampala: ACODE.

African Energy Policy Research Network/Foundation for Woodstove Dissemination (AFREPREN/FWD). (2005). *Making the African Power Sector Sustainable.* Nairobi: AFREPREN.

Alstone, P., Gershenson, D. & Kammen, D. M. (2015). Decentralized Energy Systems for Clean Electricity Access. *Nature Climate Change, 5,* 305–314. DOI: 10.1038/NCLIMATE2512.

Among, B. & Kalinaki, D. K. (2006). Government's Power Plan to Cost $4.4 Billion, *East African.* Retrieved 10 May 2006 from: http://allafrica.com/stories/printable/200605090145.html.

Ansar A., Flyvbjerg B., Budzier A. & Lunn, D. (2014). Should We Build More Large Dams? The Actual Costs of Hydropower Megaproject Development. *Energy Policy, 69,* 43–56.

Bacon, R. W. & Besant-Jones, J. (2002). *Global Electric Power Reform: Privatization and Liberalization of the Electric Power Industry in Developing Countries.* Energy and Mining Sector Board Discussion Paper Series. Paper No. 2. Washington, DC: World Bank.

Bahati, G. & Natukunda, J. F. (2009). Status of Geothermal Exploration and Development in Uganda. Paper presented at a short course on Exploration for Geothermal Resources, organized by UNU-GTP, KenGen and GDC, at Lake Naivasha, Kenya, 1–22 November 2009. Retrieved from: http://www.os.is/gogn/unu-gtp-sc/UNU-GTP-SC-10-0904.pdf.

Baker, L., Newell, P. & Phillips, J. (2014). The Political Economy of Energy Transitions: The Case of South Africa. *New Political Economy, 19(6),* 791–818.

Barkan, J. D. (2005). Uganda: An African 'Success' Past its Prime? Unpublished paper of summary remarks presented at the Woodrow Wilson International Center for Scholars.

Barkan, J. (ed.). (2009). *Legislative Power in Emerging African Democracies*. Boulder, CO: Lynne Rienner Publishers.

Barr, A., Fafchamps, M. & Owens, T. (2005). The Governance of Non-Governmental Organizations in Uganda. *World Development, 33(4)*, 657–679.

Batley, R. (2004). The Politics of Service Delivery Reform. *Development and Change, 35(1)*, 31–56.

Bayliss, K. (2002). *Privatisation and Poverty: The Distributional Impact of Utility Privatisation*. Paper No. 16. Manchester: Centre on Regulation and Competition, Institute for Development Policy and Management, University of Manchester.

Bbumba, S. (2006). Challenges Faced in Increasing Modern Energy Access: The Case of Uganda. Presentation made at World Bank Energy Week 2006 (6 March).

Birdsall, N. & Nellis, J. (2003). Winners and Losers: Assessing the Distributional Impact of Privatization. *World Development, 31(10)*, 1617–1633.

Bob, C. (2001). Marketing Rebellion: Insurgent Groups, International Media, and NGO Support. *International Politics, 38*, 311–334.

Bond, P. (n.d.). *Unsustainable South Africa: Environment, Development and Social Protest*. Scotsville, South Africa: University of Natal Press.

Bosshard, P. (2002). Private Gain – Public Risk? The International Experience with Power Purchase Agreements of Private Power Projects. Retrieved from: https://www.internationalrivers.org/resources/private-gain-public-risk-4178.

Brass, J., Carley, S., MacLean, L. M. & Baldwin, E. (2012). Power for Development: A Review of Distributed Generation Projects in the Developing World. *Annual Review of Environment and Resources, 37*, 107–136.

Bratton, M., Mattes, R. & Gyimah-Boadi, E. (2005). *Public Opinion, Democracy, and Market Reform in Africa*. Cambridge: Cambridge University Press.

Bräutigam, D. (2009). *The Dragon's Gift: The Real Story of China in Africa*. Oxford: Oxford University Press.

— (2011). China, Africa and the International Aid Architecture. In Schiere, R., Ndikumana, L. & Walkenhorst, P. (eds.). *China and Africa: An Emerging Partnership for Development?* Tunisia: African Development Bank.

— (2015). *Will Africa Feed China?* New York: Oxford University Press.

Brinkerhoff, D. W. & Crosby, B. L. (2002). *Managing Policy Reform: Concepts and Tools for Decision-Makers in Developing and Transitioning Countries*. Bloomfield, CT: Kumarian Press.

Brock, K. (2004). Ugandan Civil Society in the Policy Process: Challenging Orthodox Narratives. In Brock, K., McGee, R. & Gaventa, J. (eds.). *Unpacking Policy: Knowledge, Actors and Spaces in Poverty Reduction in Uganda and Nigeria*. Kampala: Fountain Publishers.

Brock, K., Cornwall, A. & Gaventa, J. (2001). *Power, Knowledge and Polit-*

ical Spaces in the Framing of Poverty Policy. Working Paper 143. Institute of Development Studies, University of Sussex.

Brock, K., McGee, R. & Gaventa, J. (eds.). (2004). *Unpacking Policy: Knowledge, Actors and Spaces in Poverty Reduction in Uganda and Nigeria.* Kampala: Fountain Publishers.

Brown, D. S. and Mobarak, A. M. (2009). The Transforming Power of Democracy: Regime Type and the Distribution of Electricity. *American Political Science Review, 103(2),* 193–213.

Budds, J. & McGranahan, G. (2003). Are the Debates on Water Privatization Missing the Point? Experiences from Africa, Asia and Latin America. *Environment and Urbanization, 15(2),* 87–113.

Burke, F. G. (1964). *Local Government and Politics in Uganda.* Syracuse, NY: Syracuse University Press.

Carmody, P. (2011). *The New Scramble for Africa.* Cambridge: Polity Press.

Churchill, W. (1989) [1908]. *My African Journey.* New York: W. W. Norton and Company.

Cohen, M., Madavo, C. E. & Dunkerley, H. (1983). *Learning by Doing: World Bank Lending for Urban Development, 1972–82.* Washington, DC: World Bank.

Conca, K. (2006). *Governing Water: Contentious Transnational Politics and Global Institution Building.* Cambridge, MA: MIT Press.

Daily Nation. (2005). Construction of Bujagali Power Project for 2006. *Daily Nation on the Web,* 10 May 2005, retrieved 13 July 2005 from: http://allafrica.com/stories/200505091685.html.

Daily News. (11 November 2006). Government Rules Out More Expatriates at Tanesco. Retrieved: http://www.dailynews-tsn.com/page. php?id=4437.

Davidson, O. & Karakezi, S. (1993). A New, Environmentally Sound Energy Strategy for the Development of sub-Saharan Africa. *Proceedings of the African High-level Regional Meeting on Energy and Sustainable Development.* Roskilde, Denmark: UNEP Collaborating Centre on Energy and Environment.

Davidson, O. & Mwakasonda, S. (2004). Electricity Access for the Poor: A Study of South Africa and Zimbabwe. *Energy for Sustainable Development, 8(4),* 26–40.

Davidson, O. & Sokona, Y. (2001). Energy and Sustainable Development: Key Issues for Africa. *Proceedings of the African High-level Regional Meeting on Energy and Sustainable Development.* Roskilde, Denmark: UNEP Collaborating Centre on Energy and Environment.

Day, R. & Walker, G. (2013). Household Energy Vulnerability as Assemblage. In Bickerstaff, K., Walker, G. & Bulkeley, H. (eds.). *Energy Justice in a Changing Climate.* London: Zed Books.

De Conick, J. (2004). The State, Civil Society and Development Policy in Uganda: Where Are We Coming From? In Brock, K., McGee, R. & Gaventa, J. (eds.). *Unpacking Policy: Knowledge, Actors and Spaces in Poverty Reduction in Uganda and Nigeria*. Kampala: Fountain Publishers.

DFIDa. (n.d.). *Mapping the NFP Process: An Introduction to the Series*. Uganda NFP Process Learning Series. Note 1. Retrieved from: http://www.dfid.gov.uk/pubs/files/ugandanfpseries1.pdf.

DFIDb. (n.d.). *Demonstrating Early Success: The Forest Policy Process*. Uganda NFP Process Learning Series. Note 4. Retrieved from: http://www.dfid.gov.uk/pubs/files/ugandanfpseries4.pdf.

Dicklitch, S. (2001). NGOs and Democratization in Transitional Societies: Lessons from Uganda. *International Politics, 38*, 27–46.

Dijkstra, A. G. & Kees van Donge, J. (2001). What does the 'Show Case' Show? Evidence of and Lessons from Adjustment in Uganda. *World Development, 29(5)*, 841–863.

Dowden, R. (2009). *Africa: Altered States, Ordinary Miracles*. New York: Public Affairs.

Drèze, J. & Sen, A. (1995). *India's Economic Development and Social Opportunity*. Oxford: Clarendon Press.

Eames, M. & Hunt, M. (2013). Energy Justice in Sustainability Transitions Research. In K. Bickerstaff, G. Walker & H. Bulkeley (eds.). *Energy Justice in a Changing Climate: Social Equity and Low-carbon Energy*. London: Zed Books.

East African. (30 July 2001). Power: Stick to New Tariffs, Country Told. Retrieved from: http://allafrica.com/stories/200108040036.html.

— 2005a, 22 July). Despite Extra 50 MW, Uganda Still Unable to Meet Power Demand. Retrieved from: http://allafrica.com/printable/200507220772.html.

— (2005b, 13 June). Donors to Meet 40 Percent of Uganda's Budget.

— (2006a, 14 March). WB Pulls Out of Uganda's Privatisation. Retrieved from: http://allafrica.com/stories/200603140706.html.

— (2006b, 11 July). Can the New Canadian Management Team Spark a Turnaround at Distribution Monopoly? Retrieved from: http://allafrica.com/stories/200607110093.html.

— (2007, 11 March). 'Secret' Plan to Sell Off Umeme. Retrieved from: http://allafrica.com/stories/200702200817.html.

East African Business Week. (2006a, 19 June). Uganda: $37 Million for Power. Retrieved from: http://allafrica.com/stories/200606200976.html.

— (2006b, 19 June). East Africa: Poverty, Power Top EAC Budgets. Retrieved from: http://allafrica.com/stories/200606200984.html

East African Standard. (1954a). A Varied Programme for the Queen in Uganda, p. 6.

— (1954b). Uganda's Royal Occasion, p. 4.

Easterly, W. (2013). *The Tyranny of Experts: Economists, Dictators and the Forgotten Rights of the Poor*. New York: Basic Books.

Eberhard, A. (2005). Regulation of Electricity Services in Africa: An Assessment of Current Challenges and an Exploration of New Regulatory Models. *Toward Growth and Poverty Reduction: Lessons from Private Participation in Infrastructure in sub-Saharan Africa.* Cape Town, South Africa. Retrieved from: http://www.gsb.uct.ac.za/gsbwebb/mir/documents/InfrastructureRegulationinAfrica.pdf.

Economist Intelligence Unit. (1957). *Power in Uganda: A Study of Economic Growth Prospects for Uganda with Special Reference to the Potential Demand for Electricity.* London: Economist Intelligence Unit Ltd.

Electricity Regulatory Authority. (2005). Press Release: ERA Independent Study Reveals that Ugandans Comparatively Spend Less on Electricity. Retrieved from: www.era.or.ug.

Energy for Sustainable Development. (1995). *A Study of Woody Biomass Derived Energy Supplies in Uganda.* A final report prepared for the Forest Department, Ministry of Natural Resources, the EC Financed Natural Forest Management and Conservation Project.

— (2003). *Fuel Substitution: Poverty Impacts on Biomass Fuel Suppliers.* DFID Contract No. R8019. Final Technical Report.

Engorait, S. M. (2005). Power Sector Reforms in Uganda: Meeting the Challenge of Increased Private Sector Investments and Increased Electricity Access among the Poor. In Marandu, E. & Kayo, D. (eds.). *The Regulation of the Power Sector in Africa.* London: Zed Books.

Eremu, J. (2003). Power not Social Service. *The New Vision.* Retrieved from: http://allafrica.com/stories/200312010747.html.

Ericson R. V. & Stehr, N. (eds.). (2000). *Governing Modern Societies.* Toronto: University of Toronto Press

ESMAP. (1984a). *Activity Completion Report.* Report No. 020/84.

— (1984b). *Uganda Energy Assessment.* Report No. 193/96. Washington, DC: World Bank.

— (2000). Energy Services for the World's Poor. *Energy and Development Report 2000.* Washington, DC: World Bank.

Esty, B. C. & Sesia, A. Jr. (2004). *International Rivers Network and Bujagali Dam Project (A).* Cambridge, MA: Harvard Business School. N9-204-083.

Everard, M. (2013). *The Hydropolitics of Dams: Engineering or Ecosystems?* London: Zed Books.

Fairhead, J. & Leach, M. (1993). *Misreading the African Landscape.* Tucson, AZ: University of Arizona Press.

Fallers, L. A. (1965) [1954]. *Bantu Bureaucracy.* Chicago: Chicago University Press.

Ferguson, J. (2005) [1994]. *The Anti-politics Machine: "Development," Depoliticization, and Bureaucratic Power in Lesotho.* Minneapolis, MN: University of Minnesota Press.

Fieldstone Africa. (15 November 2016). Morocco Replaces South Africa as Leading Investment Destination for Renewable Energy Projects in Africa. Retrieved from: http://www.fieldstoneafrica.com/fieldstoneafrica/news/2016/morocco-replaces-south-africa-as-leading-investment-destination-for-renewable-energy-projects-in-africa.

Fiil-Flynn, M. & Soweto Electricity Crisis Commission. (2001). *The Electricity Crisis in Soweto.* Municipal Services Project Occasional Paper No. 4. Retrieved from: http://www.municipalservicesproject.org/publication/electricity-crisis-soweto.

Flyvbjerg, B. (2003). *Megaprojects and Risk: An Anatomy of Ambition.* Cambridge: Cambridge University Press.

— (2014). What You Should Know about Megaprojects and Why: An Overview. *Project Management Journal, 45(2),* 6–19.

Forsyth, T., Leach, M. & Scoones, I. (1998). *Poverty and Environment: Priorities for Research and Policy.* Prepared for UNDP and European Commission. Retrieved from: http://eprints.lse.ac.uk/4772/.

Friedmann, J. (1992). *Empowerment: The Politics of Alternative Development.* Cambridge, MA: Blackwell.

Gaventa, J. (2004). From Policy to Power: Revisiting Actors, Knowledge and Spaces. In Brock, K., McGee, R. & Gaventa, J. (eds.). *Unpacking Policy: Knowledge, Actors and Spaces in Poverty Reduction in Uganda and Nigeria.* Kampala: Fountain Publishers.

GET FiT Uganda. (2015). *GET FiT Uganda, Annual Report 2015.* Kampala, Uganda: GET FiT Uganda.

Girod, J. & Percebois, J. (1998). Reforms in sub-Saharan Africa's Power Industries. *Energy Policy, 26(1),* 21–38.

Golooba-Mutebi, F. (2004). Reassessing Popular Participation in Uganda. *Public Administration and Development, 24,* 289–304.

Gore, C. (2009). Electricity and Privatisation in Uganda: The Origins of the Crisis and Problems with the Response. In McDonald, D. A. (ed.). *Electric Capitalism: Recolonising Africa on the Power Grid,* 359–399. Cape Town: HSRC Press.

Graham, S. & S. Marvin. (2001). *Splintering Urbanism: Networked Infrastructures, Technological Mobilities and the Urban Condition.* London: Routledge.

Grindle, M. (1980). *Politics and Policy Implementation in the Third World.* Princeton, NJ: Princeton University Press.

— (2000). *Audacious Reforms.* Baltimore, MD: Johns Hopkins University Press.

— (2004). *Despite the Odds: The Contentious Politics of Education Reform.* Princeton, NJ: Princeton University Press.

Grindle, M. & Thomas, J. (1991). *Public Choices and Policy Change: The Political Economy of Reform in Developing Countries.* Baltimore, MD: Johns Hopkins University Press.

Gunther, J. (1955). *Inside Africa.* New York: Harper & Brothers.

Hansen, H. B. & Twaddle, M. (1998). *Developing Uganda.* Oxford: James Currey.

Hardoy, J. E., Mitlin, D. & Satterthwaite, D. (2001). *Environmental Problems in an Urbanizing World.* Sterling, VA: Earthscan.

Harpham, T. & Boateng, K. A. (1997). Urban Governance in Relation to the Operation of Urban Services in Developing Countries. *Habitat International, 21(1),* 65.

Harrison, G. (2001). Post-Conditionality Politics and Administrative Reform: Reflections on the Cases of Uganda and Tanzania. *Development and Change, 32,* 657–679.

— (2004a). *The World Bank and Africa: The Construction of Governance States.* London: Routledge.

— (2004b). HIPC and the Architecture of Governance. *Review of African Political Economy, 99,* 125–173.

— (2005). The World Bank, Governance and Theories of Political Action in Africa. *BJPIR, 7,* 240–260.

Hayes, C. (1983). *Stima: An Informal History of the East Africa Power & Lighting Company.* Nairobi: East African Power and Lighting Company.

Hickey, S. (2005). The Politics of Staying Poor: Exploring the Political Space for Poverty Reduction in Uganda. *World Development, 33(6),* 995–1009.

Hill, M. (1997). *The Policy Process in the Modern State.* London: Prentice Hall.

Hira, A., Huxtable, D. & Leger, A. (2005). Deregulation and Participation: An International Survey of Participation in Electricity Regulation. *Governance, 18(1),* 53–88.

Hirschman, A. O. (1995) [1967]. *Development Projects Observed.* Washington, DC: Brookings Institution.

Human Rights Watch. (2015). 'There is no time left': Climate Change, Environmental Threats, and Human Rights in Turkana County, Kenya. Retrieved from: https://www.hrw.org/report/2015/10/15/there-no-time-left/climate-change-environmental-threats-and-human-rights-turkana.

Hyden, G. (1990). Local Governance and Economic-Demographic Transition in Rural Africa. In *Rural Development and Population: Institutions and Policy,* 193–211. New York: Oxford University Press.

— (1992). Governance and the Study of Politics. *Governance and Politics in Africa,* 1–26. Boulder, CO: Lynne Rienner Publishers.

— (2006). *African Politics in Comparative Perspective.* Cambridge: Cambridge University Press.

Hyden, G. & Bratton, M. (eds.). (1992). *Governance and Politics in Africa.* Boulder, CO: Lynne Rienner Publishers.

Hyden, G. & Court, J. (2002). Comparing Governance Across Countries and Over Time: Conceptual Challenges. *Better Governance and Public Policy: Capacity Building and Democratic Renewal in Africa,* 13–33. Bloomfield, CT: Kumarian Press.

Hyden, G., Olowu, D. & Okoth Ogendo, H. W. O. (2000). *African Perspectives on Governance.* Trenton, NJ: Africa World Press, Inc.

Infrastructure Consortium for Africa (ICA). (2010). Energy Financing Needs and Trends. Retrieved 16 December 2009 from: http://www.

icafrica.org/en/infrastructure-issues/aims10/.

Inspection Panel. (2002). *The Inspection Panel Investigation Report.* Report No. 23998. 23 May.

Inter-American Development Bank. (2003). *Keeping the Lights On: Power Sector Reform in Latin America.* Washington, DC: IADB.

— (2005). The Politics of Policies. *Ideas for Development in the Americas, Vol. 8.* Washington, DC: IADB.

International Energy Agency. (2007). *IEA Statistics 2007. Electricity Information.* IEA, Paris.

— (2011). *World Energy Outlook 2011.* Retrieved 24 January 2012 from: http://www.worldenergyoutlook.org/resources/energydevelopment/accesstoelectricity/.

International Rivers. (2010). *Africa Dams Briefing 2010.* Retrieved 5 January 2012 from: http://www.internationalrivers.org/files/attached-files/afrdamsbriefingjune2010.pdf.

— (2013). *The Downstream Impacts of Ethiopia's Gibe III Dam: East Africa's Aral Sea in the Making?* Retrieved 8 May 2013 from: http://www.internationalrivers.org/files/attached-files/impact_of_gibe_3_final_0.pdf.

— (2014). No Room for Debate on Grand Ethiopian Renaissance Dam? 17 April. Retrieved 7 May 2014 from: http://www.internationalrivers.org/blogs/229/no-room-for-debate-on-grand-ethiopian-renaissance-dam.

Isaacman, A. F. & Isaacman, B. S. (2013). *Dams, Displacement, and the Delusion of Development: Cahora Bassa and its Legacies in Mozambique, 1965–2007.* Athens, OH: Ohio University Press.

Ismail, Z. & Graham, P. (2009). *Citizens of the World? Africans, Media and Telecommunications.* Afrobarometer Briefing Paper No. 69. Retrieved from: http://afrobarometer.org/sites/default/files/publications/Briefing%20paper/AfrobriefNo69.pdf.

Iziwa, A. (2006). Sack Electricity Boss, IGG Tells Museveni. *Monitor.* Retrieved 15 July 2006 from: http://allafrica.com/stories/printable/200607130900.html.

Jacobsen, A. (2007). Connective Power: Solar Electrification and Social Change in Kenya. *World Development, 35(1),* 144–162.

Jaramogi, P. (2014). Highlights of China–Uganda Relations. *New Vision.* 5 March. Retrieved from: http://www.newvision.co.ug/new_vision/news/1338375/highlights-china-uganda-relations.

Jemaneh, Y. (2016). Public Engagement in GERD Fund Raising Intensified. *The Ethiopian Herald.* 16 April. Retrieved 3 June 2017 from: http://allafrica.com/stories/201604180732.html?aa_source=nwsltr-infrastructure-en.

Kammen, D. & Pottinger, L. (2015). Industry Insight: Hydropower: Building Sustainability into the EAPP [East African Power Pool]. *ESI Africa: Africa's Power Journal.* Retrieved 22 February 2016 from: http://www.esi-africa.com/news/industry-insight-hydropower-building-sustainability-into-the-eapp/.

Kammen, D., Jacome, V. & Avila, N. (2015). *A Clean Energy Vision for East Africa: Planning for Sustainability, Reducing Climate Risks and Increasing Energy Access.* Retrieved 22 February 2016 from: https://rael.berkeley.edu/wp-content/uploads/2015/03/Kammen-et-al-A-Clean-Energy-Vision-for-the-EAPP.pdf.
Kapika, J. & Eberhard, A. (2013). *Power Sector Reform and Regulation in Africa.* Cape Town: Human Sciences Research Council Press.
Kappel, R., Lay, J. & Steiner, S. (2005). Uganda: No More Pro-poor Growth? *Development Policy Review, 23(1),* 27–53.
Karekezi, S. & Kimani, J. (2002). Status of Power Sector Reform in Africa: Impact on the Poor. *Energy Policy, 30,* 923–945.
— (2004). Have Power Sector Reforms Increased Access to Electricity Among the Poor in East Africa? *Energy for Sustainable Development, 8(4),* 10–25.
Karekezi, S. & Mutiso, D. (2000). Power Sector Reform: A Kenyan Case Study. In Turkson, J. (ed.). *Power Sector Reform in Sub-Saharan Africa,* 83–120. Basingstoke: Macmillan Press.
Karugire, S. R. (1980). *A Political History of Uganda.* Nairobi: Heinemann Educational Books.
Kasekende, L. A. & Atingi-Ego, M. (1999). Impact of Liberalization on Key Markets in sub-Saharan Africa: The Case of Uganda. *Journal of International Development, 11,* 441–436.
Kasfir, N. & Twebaze, S. H. (2009). The Rise and Ebb of Uganda's No-Party Parliament. In Barkan, J. (ed.). *Legislative Power in Emerging African Democracies.* Boulder, CO: Lynne Rienner Publishers.
Kasita, I. (2007). World Bank Snubs Environmentalists' Calls to Delay Bujagali Power Project. Retrieved from: http://allafrica.com/stories/printable/200708010029.html.
— (2008). Debt Crunch Forced Norpak out of Karuma. 18 October. *New Vision.* Retrieved from: http://www.newvision.co.ug/new_vision/news/1178821/debt-crunch-forced-norpak-karuma.
Katorobo, J. (1996). The Public Service Reform Programme. *Uganda's Decade of Reforms 1986–1996: An Analytical Review.* Kampala: Fountain Publishers.
Keeley, J. & Scoones, I. (1999). *Understanding Environmental Policy Processes.* IDS Working Paper 89. Environment Group, Institute of Development Studies, University of Sussex.
— (2003). *Understanding Environmental Policy Processes: Cases from Africa.* London: Earthscan.
Khagram, S. (2004). *Dams and Development: Transnational Struggles for Water and Power.* Ithaca, NY: Cornell University Press.
Khunou, G. (2001). 'Massive Cutoffs': Cost Recovery and Electricity Service in Diepkloof, Soweto. *Cost Recovery and the Crisis of Service Delivery in South Africa.* London and New York: Zed Books.
Kingdon, J. (1984). *Agendas, Alternatives and Public Choices.* Boston, MA: Little, Brown.

Kiyaga-Nsubuga, J. (2004). Uganda: The Politics of 'Consolidation' under Museveni's Regime, 1996–2003. In Ali, T. & Matthews, R. (eds.). *Durable Peace: Challenges for Peacebuilding in Africa.* Toronto: University of Toronto Press.

Kjær, A. M. (2002). *The Politics of Civil Service Reform: A Comparative Analysis of Uganda and Tanzania in the 1990s.* Aarhus: Politica.

— (2004a). 'Old brooms can sweep too!' An Overview of Rulers and Public Sector Reforms in Uganda, Tanzania and Kenya. *Journal of Modern African Studies, 42(3),* 389–413.

— (2004b). *Governance.* Cambridge: Polity Press.

Komives, K., Whittington, D. & Wu, X. (2001). *Access to Utilities by the Poor: A Global Perspective.* Discussion Paper No. 2001/15. United Nations University, World Institute for Development Economics Research.

Komives, K., Whittington, D. & Wu, X. (2003). Infrastructure Coverage and the Poor: A Global Perspective. *Infrastructure for Poor People. Public Policy for Private Provision.* Washington, DC: World Bank, 77–124.

Kyokutamba, J. (2004). *Uganda: Energy Services for the Urban Poor in Africa.* London: Zed Books.

Leach, M. & Mearns, R. (1996). *The Lie of the Land: Challenging Received Wisdom on the African Environment.* Oxford: James Currey.

Leslie, J. (2005). *Deep Water: The Epic Struggle over Dams, Displaced People and the Environment.* New York: Farrar, Straus and Giroux.

Linaweaver, S. (2003). Catching the Boomerang: EIA, the World Bank, and Excess Accountability: A Case Study of the Bujagali Falls Hydropower Project, Uganda. *International Journal of Sustainable Development and World Ecology, 4,* 283–301.

Lindblom, C. (1959). The Science of Muddling Through. *Public Administration Review, 2,* 79–88.

Lofchie, M. (1989). *The Policy Factor: Agricultural Performance in Kenya and Tanzania.* Boulder, CO: Lynne Rienner Publishers.

Mackenzie, G. & Christensen, J. (1993). Tools and Methods for Energy-Environment Planning in the 1990s. In Karekezi, S. & Mackenzie, G. A. (eds.). *Energy Options for Africa: Environmentally Sustainable Alternatives.* London: Zed Books.

MacLean, L., Gore, C., Brass, J. N. & Baldwin, E. (2017.) Expectations of Power: The Politics of State-Building and Access to Electricity Provision in Ghana and Uganda. *Journal of African Political Economy & Development.* Forthcoming.

Mallaby, S. (2004). NGOs: Fighting Poverty, Hurting the Poor. *Foreign Policy,* September/October, 50–58.

Mamdani, M. (1988). Uganda in Transition: Two Years of the NRA/NRM. *Third World Quarterly, 10(3),* 1155–1181.

— (1996). *Citizen and Subject.* Princeton, NJ: Princeton University Press; London: James Currey.

Matsiko, H. (2016). Karuma, Isimba to cost extra Shs 1.8 trillion. *The Independent*. 2 May. Retrieved from: https://www.independent. co.ug/karuma-isimba-to-cost-extra-shs-1-8-trillion/3/.

Mattes, R. & Shenga, C. (2007). *'Uncritical Citizenship' in a 'Low-information Society': Mozambicans in Comparative Perspective*. Working Paper No. 91, Afrobarometer. Retrieved from: http://pdf.usaid.gov/pdf_docs/Pnadk744.pdf.

McCarney, P. (1996). Considerations on the Notion of 'Governance'. *Cities and Governance*, 3–20. Toronto: Centre for Urban and Community Studies.

— (2000). Thinking about Governance in Global and Local Perspective: Considerations on Resonance and Dissonance Between Two Discourses. *Urban Forum*, 1–38.

— (2003). Confronting Critical Disjunctures in the Governance of Cities. *Governance on the Ground: Innovations and Discontinuities in Cities of the Developing World*. Washington, DC: Woodrow Wilson Center Press.

McCarney, P. & Stren, R. (eds.). (2003). *Governance on the Ground: Innovations and Discontinuities in Cities of the Developing World*. Washington, DC: Woodrow Wilson Center Press.

McCarney, P., Halfani, M. & Rodriguez, A. (1995). Towards an Understanding of Governance. *Perspectives on the City*, 91–142. Toronto: Centre for Urban and Community Studies.

McCully, P. (2001). *Silenced Rivers: The Ecology and Politics of Large Dams*. London and New York: Zed Books.

McDonald, D. A. (2002). The Theory and Practice of Cost Recovery in South Africa. In McDonald, D. A. & Pape, J. (eds.). *Cost Recovery and the Crisis of Service Delivery in South Africa* 17–37. Cape Town and London: HSRC Publishers and Zed Books.

— (ed.). (2009). *Electric Capitalism: Recolonising Africa on the Power Grid*. Cape Town: HSRC Press.

McDonald, D. A. & Pape, J. (2002). *Cost Recovery and the Crisis of Service Delivery in South Africa*. London: Zed Books.

McGee, R. (2004). Unpacking Policy: Actors, Knowledge and Spaces. In Brock, K., McGee, R. & Gaventa, J. (eds.). *Unpacking Policy: Knowledge, Actors and Spaces in Poverty Reduction in Uganda and Tanzania*. Kampala: Fountain Publishers.

McGee, R. & Brock, K. (2001). *From Poverty Assessment to Policy Change: Processes, Actors and Data*. Working Paper 133. Institute of Development Studies, University of Sussex.

McGranahan, G. & Satterthwaite, D. (2000). Environmental Health or Ecological Sustainability? Reconciling the Brown and Green Agendas in Urban Development. *Sustainable Cities in Developing Countries*. Sterling, VA: Earthscan.

McNitt, J. R. (1982). The Geothermal Potential of East Africa. Proceedings of the Regional Seminar on Geothermal Energy in Eastern and

Southern Africa, Nairobi, Kenya, 3–8.

Mertha, A. (2008). *China's Water Warriors. Citizen Action and Policy Change.* Ithaca, NY: Cornell University Press.

Michel, S. & Beuret, M. (2009). *China Safari: On the Trail of Beijing's Expansion in Africa.* Philadelphia, PA: Nation Books.

Millán, J. & Von der Fehr, M. N. (2003). Introduction. *Keeping the Lights On: Power Sector Reform in Latin America.* Washington, DC: Inter-American Development Bank.

Ministry of Energy and Minerals. (2002a). *Energy Policy for Uganda.* Kampala, Uganda.

— (2002b). *Indicators on the Energy Policy's Successful Implementation.* Kampala, Uganda. Retrieved from: www.energyandminerals.go.ug.

Ministry of Energy and Mineral Development. (2012). *Energy Balance of Uganda.* Kampala: Government of Uganda.

— (2014). *2014 Statistical Abstract.* Kampala: Government of Uganda.

Ministry of Finance, Planning and Economic Development. (2004). *Poverty Eradication Action Plan.* Kampala: MFPED. Retrieved from: www.finance.go.ug.

Ministry of Water, Lands and Environment. (2002). *National Forest Plan.* Kampala, Uganda: MWLE.

Moehler, D. (2008). *Distrusting Democrats: Outcomes of Participatory Constitution Making.* Ann Arbor, MI: University of Michigan Press.

Moehler, D. & Singh, N. (2011). Whose News Do You Trust? Explaining Trust in Private versus Public Media in Africa. *Political Research Quarterly, 64(2),* 276–292.

Mohan, G. & Lampert, B. (2012). Negotiating China: Reinserting African Agency into China–Africa Relations. *African Affairs, 112(446),* 92–110.

Mohan, G. & Power, M. (2008). New African Choices? The Politics of Chinese Engagement. *Review of African Political Economy, 14,* 23–42.

Monitor. (2004). Tariffs Have Not Hurt Power Consumption. Retrieved from: http://allafrica.com/stories/printable/200412300052.html. 25 December 2004.

— (2005a). Construction of Bujagali Power Project for 2006. Retrieved from: http://allafrica.com/stories/200505091685.html. 13 July 2005.

— (2005b). Fire Wood to Remain Main Source of Energy. Retrieved from: http://allafrica.com/stories/printable/200502230938.html. 24 February 2005.

— (2005c). Electricity Thefts Cripple Uganda's Power Sector. Retrieved from: http://allafrica.com/stories/200502080038.html. 8 February 2005.

— (2006a). Power Crisis Deepens as Umeme Seeks New Contract. *The Monitor.* Retrieved from: http://allafrica.com/stories/200611201714. html. 22 November 2006.

— (2006b). Economy to Boom Despite Power Crisis. Retrieved from: http://allafrica.com/stories/printable/200603230725.html. 24 March 2006.

— (2006c). Ugandans Need to Talk to One Another. Retrieved from: http://allafrica.com/stories/200606210903.html.

Moore, M. (1993). Book Reviews – *Public Choices and Policy Change: The Political Economy of Reform in Developing Countries* by Merilee S. Grindle and John W. Thomas. *Journal of Development Studies, 29(4),* 269.

Mugabe, J. & Tumushabe, G. W. (1999). Environmental Governance: Conceptual and Emerging Issues. In Okoth-Ogendo, H. W. O. & Tumushabe. G. W. (eds.). *Governing the Environment: Political Change and Natural Resources Management in Eastern and Southern Africa,* 11–27. Nairobi: ACTS Press.

Mugaju, J. (2000). An Historical Background to Uganda's No-Party Democracy. In Mugaju, J. & Oloka-Onyango, J. (eds.). *No-Party Democracy in Uganda: Myths and Realities,* 8–23. Kampala: Fountain Publishers.

Mugirya, P. W. (2007). A Dam That Activists Simply Can't Make Peace With. Inter Press Service, Johannesburg. Retrieved from: http://allafrica. com/stories/printable/20070810434.html.

Mugyenzi, J. (2000). Power Sector Reform Experiences in Uganda. In Turkson, J. (ed.). *Power Sector Reform in SubSaharan Africa,* 152–175. New York: Palgrave.

Muhairwe, W. T. (2009). *Making Public Enterprises Work.* London: IWA Publishing.

Muhumuza, W. (2002). The Paradox of Pursuing Anti-Poverty Strategies under Structural Adjustment Reforms in Uganda. *The Journal of Social, Political, and Economic Studies, 3,* 271–306.

Musisi, F. (2017). World Bank to Decide on Bujagali-Isimba Dam Row. 8 February. *Monitor.* Retrieved from: http://allafrica.com/ stories/201702090028.html.

Musoke, K. D. (2001). Museveni Reduces Electricity Tariffs. *New Vision.* Retrieved from: http://allafrica.com/stories/200108280054.html.

Muwanga, D. (2008). Museveni Outlines Investment Strategy. *New Vision.* Retrieved 4 March 2016 from: http://allafrica.com/ stories/200806270208.html.

Mwenda, A. (2005). Without Mincing Words. *Monitor Online.*

Nabuguzi, E. (1995). Popular Initiatives in Service Provision in Uganda. In J. Semboja & O. Therkildsen (eds.). *Service Provision Under Stress in East Africa,* 192–208. Oxford: James Currey.

Nakaweesi, D. (2015). Why Museveni Has Made a U-turn on Privatization. *Daily Monitor.* 31 March. Retrieved from: http://www.monitor. co.ug/Business/Prosper/Why-museveni-has-made-a-u-turn-on-privatisation/-/688616/2670492/-/8rol3v/-/index.html.

Nakkazi, E. (2012). Donors Locked out of Karuma Project. *The East African.* Retrieved 4 March 2016 from: http://allafrica.com/ stories/201201231282.html.

National Research Council. (2003). *Cities Transformed: Demographic*

Change and Its Implications in the Developing World. Panel on Urban Population Dynamics, eds. Montgomery, M. R., Stren, R., Cohen, B. and Reeds, H. E. Committee on Population, Division of Behavioral and Social Sciences and Education. Washington, DC: The National Academies Press.

Nellis, J. (2003). *Privatization in Africa: What Has Happened? What Is To Be Done?* Working Paper Number 25. Washington, DC: Center for Global Development. www.cgdev.org.

Nellis, J. & Kikeri, S. (1989). Public Enterprise Reform: Privatization and the World Bank. *World Development, 17(5),* 659–672.

New Vision. (2005). Umeme Take-Over Awaits W/Bank Nod. Retrieved from: http://allafrica.com/stories/200501140477.html.

Newell, P. & Bulkeley, H. (2016). Landscape for Change? International Climate Policy and Energy Transitions: Evidence from sub-Saharan Africa. *Climate Policy,* http://dx.doi.org/10.1080/14693062.2016.1173003.

Newell, P. & Mulvaney, D. (2013). The Political Economy of the 'Just Transition'. *The Geographical Journal, 179(2),* 132–140.

Newell, P. & Phillips, J. (2016). Neoliberal Energy Transitions in the South: Kenya Experiences. *Geoforum, 74,* 39–48.

Njenga, M., Karanja, N., Karlsson, H., Jamnadass, R., Iiyama, M., Kithinji, J. & Sundberg, C. (n.d.). Additional Cooking Fuel Supply and Reduced Global Warming Potential from Recycling Charcoal Dust into Charcoal Briquette in Kenya. *Journal of Cleaner Production.* http://dx.doi.org/10.1016/j.jclepro.2014.06.002.

Njoh, A. J. (2016). A Multivariate Analysis of Inter-country Differentials in Electricity Supply as a Function of Colonialism in Africa. *Energy, 117,* 214–221.

Nordic Consulting Group. (2006). *Review of the Norwegian Support to the Energy Sector in Uganda (1997–2005).* Final Report.

Norris, P. (2001). *Digital Divide: Civic Engagement, Information Poverty, and the Internet Worldwide.* Cambridge: Cambridge University Press.

Odyek, J., Musoke, C., Birungi, A. & Tatyama, A. (2001). Power Tariffs to Stay. *New Vision.* 15 August. Retrieved from: http://allafrica.com/stories/200108150331.html.

Okudo, I. (2016). Govt Officials Fight Over Isimba, Karuma. *Daily Monitor.* (30 March). Retrieved from: http://mobile.monitor.co.ug/News/Govt-officials-fight-over-Isimba--Karuma/2466686-3138634-format-xhtml-266v18z/index.html.

Okwello, J. (2002). Museveni Warns Economic Saboteurs. *New Vision.* Retrieved from: http://allafrica.com/stories/200201270092.html. 27 January 2002.

Olowu, D. (2002). Governance, Institutional Reforms and Policy Processes in Africa: Research and Capacity-Building Implications. *Better Governance and Public Policy: Capacity Building and Public Policy,* 53–71. Bloomfield, CT: Kumarian Press.

Olowu, D. & Sako, S. (2002). *Better Governance and Public Policy: Capacity Building and Public Policy.* Bloomfield, CT: Kumarian Press.

Olukoju, A. (2004). 'Never Expect Power Always': The Electricity Consumers' Response To Monopoly, Corruption and Inefficient Services in Nigeria. *African Affairs, 103,* 51–71.

Ondaatje, C. (1998). *Journey to the Source of the Nile.* Toronto: Harper-Collins.

Onyango-Obbo, C. (2006). 'In Generator City, Birds Don't Sing, And You Have No E-Mail', *East African.* Retrieved from: http://allafrica.com/stories/200605300449.html. 30 May 2006.

Ostrom, E., Wynne, S. & Schroeder, L. (eds.). (1993). *Institutional Incentives and Sustainable Development: Infrastructure Policies In Perspective.* Boulder, CO: Westview Press.

Park, S. (2005). Norm Diffusion within International Organizations: A Case Study of the World Bank. *Journal of International Relations and Development, 8,* 111–141.

— (2009). The World Bank, Dams and the Meaning of Sustainable Development in Use. *Journal of International Law and International Relations, 5(1),* 93–122.

Peters, B. G. (2000). Governance and Comparative Politics. *Debating Governance: Authority, Steering and Democracy,* 36–53. Oxford: Oxford University Press.

Pierre, J. (2000). Introduction: Understanding Governance. *Debating Governance: Authority, Steering and Democracy,* 1–10. Oxford: Oxford University Press.

Pineau, P. (2002). Electricity Sector Reform in Cameroon: Is Privatization the Solution? *Energy Policy, 30(11/12),* 999–1012.

Prayas Energy Group. (2002). *The Bujagali Power Purchase Agreement – An Independent Review.* A Study of Techno-Economic Aspects of the Power Purchase Agreement of the Bujagali Hydroelectric Project in Uganda. Report Prepared for International Rivers Network.

Pierson, P. (2000). Increasing Returns, Path Dependence, and the Study of Politics. *American Political Science Review, 94(2),* 251–267.

Power, M., Newell, P., Baker, L., Bulkeley, H., Kirshner, J. & Smith, A. (2016). The Political Economy of Energy Transitions in Mozambique and South Africa: The Role of the Rising Powers. *Energy Research & Social Science, 17,* 10–19.

Pritchard, S. B. (2012). From Hydroimperialism to Hydrocapitalists: 'French' Hydraulics in France, North Africa, and Beyond. *Social Studies of Science, 42(4),* 591–615.

Ranganathan, V. (1998). The Power Generation Sector in Africa: An Overview. *Planning and Management in the African Power Sector,* 3–14. Chapter contributions by: Lucy Khalema-Redeby, Hailu Mariam, Abel Mbewe and Ben Ramasedi. New York: Zed Books.

Reinikka, R. & Svensson, J. (2001). The Cost of Doing Business: Ugandan Firms' Experiences with Corruption. *Uganda's Recovery: The Role of Farms, Firms, and Government.* Washington, DC: The World Bank.

Republic of South Africa. (1998). *White Paper on Energy Policy for the Republic of South Africa.* Department of Minerals and Energy.

Roe, E. (1991). Development Narratives, Or Making the Best of Blueprint Development. *World Development, 19(5),* 287–300.

— (1994). *Narrative Policy Analysis.* Durham, NC: Duke University Press.

Rotberg, R. (2013). *Africa Emerges: Consummate Challenges, Abundant Opportunities.* Cambridge: Polity Press.

Rycroft, R. W. & Szyliowicz, J. S. (1980). The Technological Dimension of Decision Making: The Case of the Aswan High Dam. *World Politics, 33(1),* 36–61.

Saghir, J. (2005). *Energy and Poverty: Myths, Links and Policy Issues.* Energy Working Notes, Energy and Mining Sector Board, No. 4, May. Washington, DC: World Bank.

Schnurr, M. & Gore, C. (2015). Getting to 'Yes': Governing Genetically Modified Crops in Uganda. *Journal of International Development, 27(1),* 55–72.

Scoones, I., Leach, M. & Newell, P. (2015a). *The Politics of Green Transformations.* New York: Routledge.

Scoones, I., Newell, P. & Leach, M. (2015b). The Politics of Green Transformation. In Scoones, I., Leach, M. & Newell, P. (eds.). *The Politics of Green Transformation,* 1–24. New York: Routledge.

Scott, J. (1998). *Seeing Like a State.* New Haven, CT: Yale University Press.

Sen, A. (1999). *Development as Freedom.* New York: Anchor.

Sovacool, B. K. & Dworkin, M. H. (2014). *Global Energy Justice: Problems, Principles and Practices.* Cambridge: Cambridge University Press.

Speke, J. H. (1967) [1864]. *What Led to the Discovery of the Source of the Nile.* London: Frank Cass.

Ssewakiryanga, R. (2004). The Corporatist State, the Parallel State, and Prospects for Representative and Accountable Policy. In Brock, K., McGee, R. & Gaventa, J. (eds.). *Unpacking Policy: Knowledge, Actors and Spaces in Poverty Reduction in Uganda and Nigeria.* Kampala: Fountain Publishers.

Stehr, N. & Ericson, R. V. (2000). The Ungovernability of Modern Societies: States, Democracies, Markets, Participation, and Citizens. *Governing Modern Societies,* 3–25. Toronto: University of Toronto Press.

Stiglitz, J. E. (2002). *Globalization and its Discontents.* New York, London: Norton.

Stren, R. E. (1992). African Urban Research since the Late 1980s: Responses to Poverty and Urban Growth. *Urban Studies, 29(3/4),* 533–555.

— (2003). Introduction: Toward the Comparative Study of Urban

Governance. In McCarney, P. & Stren, R. E. (eds.). *Governance on the Ground* 1–30. Washington, DC: Woodrow Wilson Center Press.

Stren, R. E. & Polese, M. (eds.). (2000). *The Social Sustainability of Cities.* Toronto: University of Toronto Press.

Talemwa, M. (2012). Power Crisis Weighs Down Economy. *The Observer.* 19 January. Retrieved from: http://allafrica.com/stories/201201200296.html.

Tangri, R. & Mwenda, A. (2001). Corruption and Cronyism in Uganda's Privatization in the 1990s. *African Affairs, 100(398)*, 117–133.

Taylor, I. (2010). *China's New Role in Africa.* Boulder, CO: Lynne Rienner Publishers.

Tendler, J. (1965). Technology and Economic Development: The Case of Hydro vs. Thermal Power. *Political Science Quarterly, 80(2)*, June, 236–253.

— (1968). *Electric Power in Brazil.* Boston, MA: Harvard University Press.

— (1979). *Rural Electrification: Linkages and Justifications.* A.I.D. Program Evaluation, Discussion Paper No. 3. Retrieved from: http://www.mit.edu/~tendler/fs/works/Rural%20Electrification%20.pdf.

Tenywa, G. (2003). KCC Protests Cutting Down of Trees. *New Vision.* 18 January. Retrieved from: www.newvision.co.ug.

Therkildsen, O. (2000). Public Sector Reform in a Poor, Aid-dependent Country, Tanzania. *Public Administration and Development, 20*, 61–71.

Tripp, A. M. (2000). *Women and Politics in Uganda.* Madison, WI: University of Wisconsin Press; Oxford: James Currey.

— (2004). The Changing Face of Authoritarianism in Africa: The Case of Uganda. *Africa Today, 50(3)*, 3–26.

— (2010). *Museveni's Uganda: Paradoxes of Power in a Hybrid Regime.* Boulder, CO, London: Lynne Rienner Publishers.

Tukahebwa, G. B. (1998). Privatization as a Development Policy. *Developing Uganda*, 59–72. Oxford: James Currey.

Turkson, J. & Wohlgemuth, N. (2001). Power Sector Reform and Distributed Generation in sub-Saharan Africa. *Energy Policy, 29*, 135–145.

Twaddle, M. & Hansen, H. B. (1998). The Changing State of Uganda. In Hansen, H. B. & Twaddle, M. *Developing Uganda*, 1–18. Oxford: James Currey.

Uganda. (1998). *Hansard Parliamentary Debates*, 4732–4733 (Mr Mutyaba MP).

Uganda Electricity Board. (1996). *The Thirty Third Report and Accounts of Uganda Electricity Board.* Kampala: Uganda Electricity Board.

— (1999). *Report and Accounts of 1999.* Kampala: Uganda Electricity Board.

UNCHS. (2001). *Cities in a Globalizing World: Global Report on Human Settlements 2001.* London: Earthscan.

USAID. (2005). Uganda. Retrieved from: http://www.usaid.gov/policy/

budget/cbj2005/afr/ug.html. 28 November 2005.
Van De Walle, N. (1989). Privatization in Developing Countries: A
Review of the Issues. *World Development, 17(5)*, 601–615.
Vedavalli, R. (2007). *Energy for Development: Twenty-First Century
Challenges of Reform and Liberalization in Developing Countries.*
London: Anthem Press.
Verhoeven, H. (2011). 'Dams are Development': China, the Al-Ingaz
Regime and the Political Economy of the Sudanese Nile. In Large, D.
& Patey, L. A. (eds.). *Sudan Looks East. China, India and the Politics
of Asian* Alternatives, 120–138. Oxford: James Currey.
— (2015). *Water, Civilisation and Power in Sudan: The Political Economy
of Military-Islamist State-Building.* Cambridge: Cambridge Univer-
sity Press.
Wakabi, W. (2004). Umeme to Manage Uganda's Power Supply. *East
African.* Retrieved from: http://allafrica.com/stories/200408030882.
htm.
Wakabi, M. (2013). Museveni's Battle with his Bureaucracy. *East Africa*,
p. 1, 6.
Wamukonya, N. (2003). Power Sector Reform in Developing Countries:
Mismatched Agendas. *Energy Policy, 31*, 1273–1289.
Wereko-Brobby, C. (1993). Innovative Energy Policy Instruments
and Institutional Reform: The Case of Ghana. In Karekezi, S. and
Mackenzie, G. A. (eds.). *Energy Options for Africa.* London: Zed
Books.
Wilson, G. (1967). *Owen Falls: Electricity in a Developing Country.*
Nairobi: East African Publishing House.
World Bank. (1993). *The World Bank's Role in the Electric Power Sector.*
A World Bank Policy Paper. Washington, DC: World Bank.
— (2000a). *Fuel for Thought: An Environmental Strategy for the Energy
Sector.* Washington, DC: World Bank.
— (2000b). *Uganda – Fourth Power Project. Project Information Docu-
ment (PID).* Report No. UGPE2984. Washington, DC: Africa Regional
Office.
— (2000c). *Reforming Institutions and Strengthening Governance: A
World Bank Strategy.* Washington, DC: World Bank.
— (2001a). *Bujagali Hydropower Project. Project Information Document
(PID).* Report No. PID8803. Washington, DC: World Bank.
— (2001b). *Uganda: Policy, Participation, People.* Operations Evaluation
Department. Washington, DC: World Bank.
— (2002). *Sustainability in the Electricity Utility Sector.* Retrieved from:
www.wbcsd.ch.
— (2003). *Infrastructure for Poor People: Public Policy for Private Provision.* Washington, DC: World Bank.
— (2004a). *Water Resources Sector Strategy: Strategic Directions for
World Bank Engagement.* Washington, DC: World Bank.
— (2004b). *Reforming Infrastructure: Privatization, Regulation and*

Competition. Washington, DC: World Bank and Oxford University Press.

— (2005a). *Uganda Country Brief.* 5 May. Retrieved from: http://web.worldbank.org/WBSITE/EXTERNAL/COUNTRIES/AFRICAEXT/UGANDAEXTN/0,,menuPK:374947~pageP-K:141132~piPK:141107~theSitePK:374864,00.html.

— (2005b). *Project Completion Note (Guarantee No. B-003-0-UG).* Report No. 33722-UG. Washington, DC: World Bank.

— (2008). *The Welfare Impact of Rural Electrification: A Reassessment of the Costs and Benefits.* Washington, DC: World Bank Independent Evaluation Group.

— (2009). *Directions in Hydropower.* Washington, DC; World Bank. Retrieved 14 January 2013 from: http://water.worldbank.org/publications/directions-hydropower. World Business Council for Sustainable Development.

— (2010). *The Welfare Impact of Rural Electrification: A Reassessment of Costs and Benefits.* Washington, DC: World Bank Impact Evaluation Group.

— (2013). Learning by Doing: The Social Observatory. 20 March. Retrieved from: http://www.worldbank.org/en/news/feature/2013/03/20/learning-by-doing-social-observatory.

— (2015a). Kenya's Geothermal Investments Contribute to Green Energy Growth, Competitiveness and Shared Prosperity. Retrieved 13 January 2016 from: http://www.worldbank.org/en/news/feature/2015/02/23/kenyas-geothermal-investments-contribute-to-green-energy-growth-competitiveness-and-shared-prosperity.

— (2015b). *Bujagali Indemnity Agreement with the Government of Uganda and the Proposed Isimba Hydropower Project – Fact Sheet.* 15 July. Retrieved from: http://www.worldbank.org/en/country/uganda/brief/bujagali-indemnity-agreement-with-the-government-of-uganda-and-the-proposed-isimba-hydropower-project-fact-sheet.

World Bank and Energy Sector Management Assistance Programme (ESMAP). (2000). *Energy Services for the World's Poor.* Washington, DC: World Bank.

World Commission on Dams. (2000). *Dams and Development: A New Framework for Decision-making.* London and Sterling, VA: Earthscan.

World Resources Institute. (2000). *Power Politics: Equity and Environment in Electricity Reform.* Washington, DC: World Resources Institute.

Yi-chong, X. (2006). The Myth of Single Solution: Electricity Reforms and the World Bank. *Energy, 31,* 802–814.

INDEX